D1217606

GOD'S EQUATION

GOD'S EQUATION

EINSTEIN, RELATIVITY AND THE EXPANDING UNIVERSE

AMIR ACZEL

Published in the UK in 2000 by
Judy Piatkus (Publishers) Limited
5 Windmill Street
London W1P 1HF
e-mail: *info@piatkus.co.uk*

For the latest news and information on all our titles,
visit our website at www.piatkus.co.uk

First published in 1999 by
Four Walls Eight Windows, New York, USA

Albert Einstein™ licensed by the Hebrew University of Jerusalem.

A catalogue record for this book is available from the British Library

ISBN 0 7499 2082 3

Printed and bound in Great Britain by
Butler & Tanner Ltd, Frome, Somerset

To my father, Captain E. L. Aczel

Contents

Preface <ix>

CHAPTER 1
Exploding Stars <1>

CHAPTER 2
Early Einstein <13>

CHAPTER 3
Prague, 1911 <27>

CHAPTER 4
Euclid's Riddle <43>

CHAPTER 5
Grossmann's Notebooks <61>

CHAPTER 6
The Crimean Expedition <71>

CHAPTER 7
Riemann's Metric <91>

CHAPTER 8
Berlin <105>

CHAPTER 9
Principe Island <121>

CHAPTER 10
The Joint Meeting <139>

CHAPTER 11
Cosmological Considerations <149>

CHAPTER 12
The Expansion of Space <167>

CHAPTER 13
The Nature of Matter <181>

CHAPTER 14
The Geometry of the Universe <189>

CHAPTER 15
Batavia, Illinois, May 4, 1998 <197>

CHAPTER 16
God's Equation <207>

References <221>
Index <225>

Sehr geehrter Herr Kollege

Unter k verstehe ich die Konstante, die so ins Newton'sche Gesetz eingeht

$$\text{Kraft} = k\,\frac{m\,m'}{r^2}$$

Das Masssystem ist in der Formel

$$\alpha = \frac{2kM}{c^2\Delta}$$

natürlich belanglos, wenn nur c im gleichen System gemessen wird.

Herrn
Dr. E. F. Freundlich
Kgl. Sternwarte Berlin S.W. 48
Enckeplatz 3a
Berlin

Card dated September 21, 1911, from Albert Einstein to Erwin Freundlich.
The Pierpont Morgan Library, New York, MA 4725.

Preface

In January 1998, the way we perceive the universe changed forever. Astronomers found evidence that the cosmos is expanding at an ever-increasing rate. As soon as the new findings were announced, cosmologists from all over the world rushed to try to explain the underlying phenomenon. The most promising theory these scientists could come up with was one that Albert Einstein had proposed eight decades earlier and quickly retracted, calling it his greatest blunder. Every year, new developments prove the accuracy of Einstein's theories. But if the cosmologists' new assessments are correct, then Einstein was right even when he was sure he was wrong.

About the time this astounding news was being reported, I received a curious piece of mail. It was sent to me by L. P. Lebel, a reader of my book *Fermat's Last Theorem* who had become a friend, and with whom I'd been exchanging letters. This time, however, the envelope contained no letter: simply a cutting of an article written by George Johnson in the *New York Times*. I read the article with great interest: it was about pure mathematics, not physics or cosmology. In the article, Mr. Johnson posed an intriguing question: Is it possible that other forms of mathematics—different from our own—exist somewhere in the universe? As an example, Johnson gave the problem of pi and the circle. Is it possible, he asked, for circles to exist with ratio of circumference to diameter not equal to pi?

Einstein and cosmology would seem, on the face of it, to have nothing to do with weird mathematics where circles are not the way we know them. But a strong connection existed, as I knew well. Contemplating these parallel issues of physics and mathematics took me back two decades. During my student days at the University of California at Berkeley, I took courses in physics and mathematics. In one of these courses, the professor explained a concept that challenged my perception of things. "The electron," the professor said, "lives in a different space from the one we live in." This statement made me change my academic direction, and from then on pursue courses that dealt with spaces: topology, analysis, and differential geometry. I wanted to understand these different spaces that exist even if our senses cannot detect them. Such strange spaces apply to the very small (in quantum mechanics) or the very large (in general relativity). To understand the physics of relativity, one had to study a space whose geometry worked counter to our intuition.

Johnson's weird mathematics and the cosmologists' Einsteinian equations were really two sides of the same coin. Slowly, I became obsessed with these fascinating ideas. I spent hours solving problems in non-Euclidean geometry, a branch of mathematics that deals with spaces in which a line may have infinite parallels through a given point instead of Euclid's single line, and where circles have ratios of circumference to diameter different from pi. (Albert Einstein studied non-Euclidean geometry when he was searching for a mathematical theory that would account for the curvature he discovered in space-time.) I reworked old problems in differential geometry—another form of geometry used by Einstein when he needed a mathematical basis for his nascent theory of general relativity. And I spent time going through all of Einstein's papers on general relativity.

Recently, having refreshed my understanding of the mathematics of relativity theory, I called one of my old Berkeley professors to ask him some questions about the geometry of general relativity. S. S. Chern is arguably the greatest living geometer. We spoke on the phone for a long time, and he patiently answered all my questions. When I told him I was contemplating writing a book about relativity, cosmology, and geometry and how they interconnect to explain the universe, he said: "It's a wonderful idea for a book, but writing it will surely take too many years of your life . . . I wouldn't do it." Then he hung up.

I was determined to explain to *myself* the exact relationship between an ever-expanding universe, Einstein's ingenious field equation of general relativity, and the enigmatically curved universe in which we live. If I could explain these mysteries to myself and satisfy my own growing curiosity, I felt, then I could share this knowledge with others. I read every book I could find about cosmology and relativity, but to truly understand these fascinating ideas, I had to derive the equations myself. In this task, other people proved more helpful than I would ever have expected.

My friend and neighbor Alan Guth, the Weisskopf Professor of Physics at MIT, is the discoverer of the most promising theory to explain what happened right after the big bang—the theory of the inflationary universe. Guth's theory is so successful that it is now the backbone of virtually every cosmological model of the universe. Alan was generous in sharing with me his research papers and spending hours with me discussing cosmology and the strange geometry of space-time. Peter Dourmashkin, who also teaches physics at MIT, kindly shared his lecture notes on cosmology and helped me through some hairy equations. Jeff Weeks, a mathematician and consultant, helped me see the exact mathematical connection between Einstein's field equation with a cosmological constant and the possible geometries of the universe. Colin Adams, a mathematician at Williams College, was helpful in further exploring these hidden links between geometry and the mathematical formulas describing the universe. Kip Thorne, a world-renowned professor of relativity at Caltech and an expert on black holes, was kind enough to answer questions in a telephone interview. Paul Steinhardt, a professor of physics at Princeton University and a pioneer in the areas of cosmology and physics as well as pure mathematics, shared his insights and theories with me. Sir Roger Penrose of Oxford University, a renowned mathematician and cosmologist, generously shared his original ideas and theories of the universe.

Once I understood the mathematics and the physics and could actually see how the equations determined the geometry and how Einstein's once-maligned cosmological constant amazingly fit into the puzzle of an ever-accelerating universe, it came time to speak with the astronomers: harbingers of news about the state of our universe. Saul Perlmutter, of the Lawrence Berkeley National Laboratory, head of the international team of astronomers who broke the news about the fast-expanding universe, was generous with his time. Saul gave me unique insights into the physical process of an expanding space, as well as information on the ingenious methods he and his team members had devised in order to detect and measure universal expansion using electronic images of exploding stars billions of light years away. Later, Saul also went through the manuscript of this book and gave me many valuable suggestions. Esther M. Hu, head of a team of astronomers at the University of Hawaii who glimpsed through the Keck telescopes a vision of the most distant object visible in the universe—a galaxy 13 billion light years away whose light was so dim and so far shifted to red as to seem at the limit of what we could hope to see—described to me her amazing experience. She was also generous in providing me with many interesting technical details about her discovery, including the fact that the galaxy she observed was receding from us at 95.6 percent of the speed of light. Neta A. Bahcall, a professor of astronomy at Princeton University who has been studying the mass density of the universe with the most advanced observational and theoretical tools, shared with me her surprising research results. All the studies Neta and her colleagues have conducted over the past decade indicate that our universe has low mass—as little as 20 percent of the minimum mass density required to eventually stop the universal expansion. This research gives strong indications that the universe will expand forever.

My friend Jay Pasachoff, director of the Hopkins Observatory at Williams College, was my host at Williamstown, Massachusetts, one summer day while I was deep into the project of writing this book. I had come to talk with him because I was now researching the work of Albert Einstein himself. I knew that Einstein's theory of general relativity was confirmed from observations of the bending of starlight around the sun during the total solar eclipse of 1919. Jay Pasachoff is the world's leading authority on solar eclipses. By the time of our meeting he had observed 26 total solar eclipses, which I believe is more eclipses than any one person has ever observed in the history of this planet. Since then, Jay has seen even more eclipses. Jay gave me entire files of original documents and articles. Then he gave me an article about a collection of letters Albert Einstein wrote over a period of 20 years to an obscure German astronomer. These letters had just been donated by a private collector to the Pierpont Morgan Library in Manhattan. Many had never been seen by scholars, nor translated. I knew that there was a good story here to be discovered.

Sylvie Merian and Inge Dupont of the Pierpont Morgan Library were most helpful to me over the several hours I spent at the archives room at the library looking at the letters Einstein wrote to the astronomer Erwin Freundlich. They kindly supplied me with official copies of all 25 of Einstein's letters in the collection. I thank Charles Hadlock for help in arranging the visit.

My father, Captain E. L. Aczel, was spending the summer with us in Boston. My father was raised and educated in the Austro-Hungarian Empire—before leaving in the 1930s to become a ship's captain in the Mediterranean—and is an expert on the kind of German that Albert Einstein spoke and wrote during those same years. When I asked him if he would mind spending some of his time translating Einstein's letters, my father was delighted to comply. Over the next two months, we spent many long hours together working on the letters. He would often go back to a sentence or phrase after we had finished translating a letter, and ponder a spicy expression Einstein had used ("Your nerves are frayed and you don't even have a coating of bacon on your head to protect you."), or what the physicist really thought when he brushed off his young colleague's request for help with his job ("Struve cursed you today. You don't do what he tells you to do."). My father's careful eye and ear, his attention to every minute detail and its meaning within the dialect of the time and place, revealed a surprising new picture of Albert Einstein. The kindly old man famous for his humanity was still there, but it was clear that Einstein was not only extremely ambitious, he was ready to use people to achieve his goals and to drop them quickly once they were no longer useful to him. The legendary physicist seemed more human now—with the flaws present in all of us.

On my visit to the Einstein Archives in Jerusalem I saw the other side of the Freundlich-Einstein relationship as reflected in the letters Freundlich wrote to Einstein. I am indebted to Dina Carter of the Albert Einstein Archives at the Jewish National and University Library in Jerusalem for guiding me to many important letters and documents.

People who spend their careers studying the life and work of Albert Einstein form a tight-knit international community, even as they are spread over the globe from Boston to Princeton to Zürich, Jerusalem, and Berlin. John Stachel of Boston University, the founding editor of the many volumes of the *Collected Papers of Albert Einstein*, provided me with useful information on the chronology of some of Einstein's discoveries. My friend Hans Künsch of the ETH in Zürich, the Swiss polytechnic school at which Einstein studied and taught, made arrangements for me to see Einstein's house in Switzerland.

At the Max Planck Institute for the History of Science in Berlin I met two of the world's greatest experts on the work of Albert Einstein. Jürgen Renn, the director of the institute, postponed his departure for a vacation on an island in the Baltic Sea so he could meet with me during my visit to Berlin. Renn and his colleagues at the institute have discovered many facts about the science of Albert Einstein, including the stunning finding that Einstein actually had written in his notebook the exact form of his final field equation of gravitation as early as 1912, only to discard it for unknown reasons and rediscover the same formula after four more years of hard work—developing the equation from another angle. Jürgen made available to me the resources of his institute, allowing me to inspect many yet-unpublished findings about Einstein and his work. Giuseppe Castagnetti, a

fellow researcher at the Max Planck Institute, also helped me a great deal during my stay in Berlin. I am grateful to him for many insights about Einstein's personality and work. Giuseppe also arranged for me to see Einstein's country house in Caputh.

While in Berlin, I was disappointed to find that neither of Einstein's residences in the city, Wittelsbacherstrasse 13 and Haberlandstrasse 5, are marked. The Berlin authorities mark with a commemorative plaque the location where every minor official of the government or a minor poet or artist spent even a few months—but not the long-term residences of the greatest physicist of all time. I found this puzzling and somewhat disturbing. I made a mental note of the fact that both unmarked Einstein residences were in the part of the city that was West Berlin. At the building that housed the Prussian Academy of Sciences—on Unter den Linden Avenue in the heart of what used to be East Berlin—there was, indeed, a small plaque commemorating the great scientist's tenure at the academy from 1914 to 1932.

In Caputh (a village in what used to be known as East Germany), a surprise awaited me. Not only is Einstein's house well-marked, but the building is maintained as a monument to the great scientist, and groups of visitors are brought in to visit it. I am grateful to Mrs. Erika Britzke, who maintains the house, for giving me a special tour of the facilities, including parts of the house not open to the public, and for sharing with me her wealth of information about the Einstein family and the time they spent at the house.

In late summer 1998, after much of my research leading to this book was completed and I felt that I was able to tie together the cosmological theories, the astronomical discoveries, the physics of gravity and spacetime, and Einstein's personal odyssey of discovery, I had a visitor. My good friend, Carlo F.

Barenghi, a physicist and mathematician at the University of Newcastle, England, came to stay with us. Carlo was attending a conference on quantum theory in the Berkshire Mountains of western Massachusetts, and every night he and I drove back to Boston together. We spent the time in the car talking about cosmology and the riddle of the universe. Carlo helped me sharpen some of the cosmological arguments in this book.

I thank my publisher, John Oakes, for his support and encouragement, and I thank the dedicated staff at Four Walls Eight Windows in New York: Kathryn Belden, Philip Jauch, and JillEllyn Riley.

My wife, Debra Gross Aczel, who teaches writing at MIT, read through the entire manuscript and offered me many suggestions for improving the book. I am grateful to you, Debra, for everything you have done, and I am grateful to all the wonderful people mentioned in this preface for their enthusiasm, help, information, and suggestions.

CHAPTER 1

Exploding Stars

"What is so remarkable is that we are answering deep philosophical questions with physical measurements."
—Saul Perlmutter

Saul Perlmutter sat in his office high in the Berkeley hills overlooking the San Francisco Bay and watched the sun set below the Golden Gate Bridge. It was a magnificent sight, the sun turning redder and its shape layering into rectangular pieces that then slowly disappeared into the blue-gray Pacific Ocean beyond. He knew why sunsets were red and why the sky was blue—Saul Perlmutter is an astrophysicist. And it was exactly this phenomenon, so common on Earth and watched by millions of people every day from hilltops or beaches or restaurants on top of skyscrapers, that was puzzling him now because of its implication about the exploding stars he had seen halfway across the universe.

For ten years, Saul Perlmutter has directed an effort by a team of astronomers from his headquarters at the Lawrence Berkeley National Laboratory on the hills across the Golden Gate. Using the most advanced telescopes, in Hawaii, Chile, and in space, the astronomers have been collecting electronic images of distant galaxies, many thousands of galaxies at a time, and comparing the images with those of the same galaxies taken three weeks later. The astronomers are looking for exploding stars within

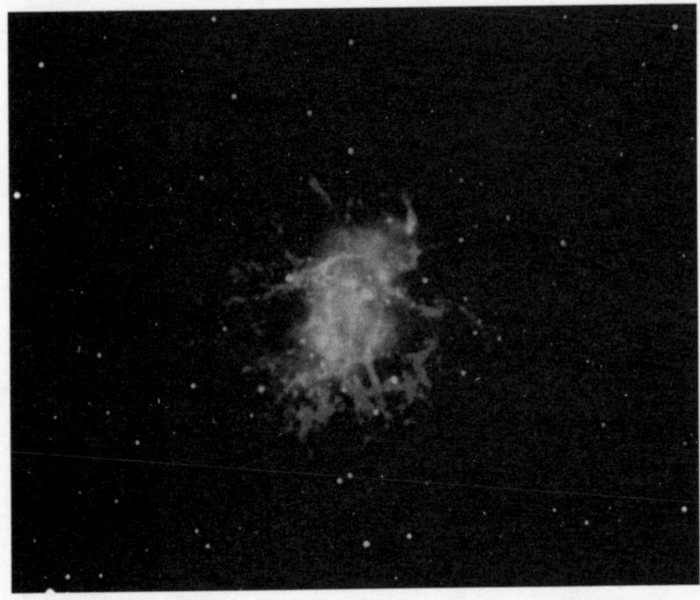

M1: Crab Nebula
T. Credner & S. Kohle, Bonn University, Calar Alto Observatory

these very distant galaxies. An explosion appears as a relatively bright spot of light on the photograph (actually, an electronic image) of the galaxy, and is absent on the image taken three weeks earlier. These scientists were not searching for ordinary explosions. They were looking for Type Ia supernovae—among the most gargantuan explosions ever observed in the cosmos.

In AD 1054, Chinese astronomers recorded a "guest star" that had appeared suddenly in the vicinity of the star we know today as Zeta Tauri—the tip of one of the long horns of Taurus, the Bull. Within a month, the star disappeared, but a nebula remained, visible today with a medium-power telescope. This faint, cloudlike object is denoted M1, or the Crab Nebula, because of the shape of the faint formation. The Crab Nebula is a giant cloud of gas and dust that remained from the explosion of an ancient star and has been expanding into the surrounding space ever since. In the center of the nebula lives the star's collapsed core—a neutron star, pulsating with intense radiation beamed into space every fraction of a second—a *pulsar*. The "guest star" was not a star at all. What the Chinese had observed was the intense light from the explosion of a star so far away that it could not be seen before the bright explosion. Such an explosion is called a *supernova*.

The word *nova* means new, and a "nova"—a sudden brightening of an invisible star—was thought to be the birth of a new star. Such sudden brightening occurs when a white dwarf (a form of dead star) attracts matter from an orbiting companion and brightens to a level that makes it briefly visible. A supernova is a much brighter occurrence, and we know it to be caused by the explosion of a star. Ironically, it signifies the death of a star, rather than its birth. In 1987, a supernova was observed in the southern hemisphere by modern-day astronomers, and the wealth of results from their research has taught us much about these mysterious explosions in the night sky. Supernovae have been observed by astronomers over the last three centuries, but the 1987 explosion was the first that could be seen with the naked eye. This was a Type II.

When a massive star—much more massive than the Sun—is through converting its hydrogen into helium and helium into carbon, and the later nuclear reactions that make it burn as a bright star have all been exhausted, the star can no longer hold itself up against gravitational collapse. As it falls inward under its own weight, the star explodes spectacularly. This explosion is called a Type II supernova. Then, depending on its size, the star's remnants will turn into a dense dead body called a neutron star (in which ordinary protons and electrons can no longer coexist and they fuse together to form neutrons), or—in the more massive cases—a black hole, the most bizarre object in the universe. In this latter case, the object is so dense and its gravitational pull so immense that even light cannot escape it.

But the supernovae that Saul Perlmutter and his team were studying in their quest to understand the universe were something totally different. These explosions could rightly be called super-supernovae, although scientists simply call them Type Ia supernovae. A Type Ia supernova is six times as bright as an "ordinary" supernova. In the visible range of radiation, such an explosion is the brightest phenomenon ever observed in space. A Type Ia supernova occurs after a white dwarf, the dead remains of a star of the same type as our Sun (which will, itself, become a white dwarf when it is through with its own nuclear fuels in another five billion years), begins to collect matter that falls into it from a nearby companion star, each star orbiting the other. Once the incoming matter inflates the mass of the white dwarf to about 1.4 times the mass of our Sun, a sudden explosion of unequaled violence occurs. In this type of supernova, the matter blown into space from the exploding white dwarf can reach speeds that are a measurable fraction of the speed of light.

The brightness of the Type Ia supernova makes it almost as luminous as an entire galaxy. The explosion is immense—and clearly identifiable by its characteristics. And because of this latter property, finding such supernovae has become an urgent goal for all astronomers interested in measuring the distance and speed of recession of faraway galaxies. These exploding stars are like beacons in the sky. Their relative brightness, the observed brightness as a fraction of the brightness one would expect if the explosions were nearby (within our own galaxy), can tell astronomers how far the stars' home galaxies are from Earth.

Astronomers can also estimate the speed of recession of the distant galaxies by measuring their redshifts. A redshift is the increase in wavelength that a ray of light undergoes when its source is receding from the observer. The principle of this phenomenon is the Doppler effect familiar from everyday life—the change in pitch that a sound wave undergoes when, for example, a speeding train passes the observer. In light, a similar change in frequency takes place: the wavelength of light rays increase, that is, move toward the red end of the spectrum, when the source recedes from the observer; and wavelengths decrease, that is, shift toward the blue end of the spectrum, when the light source approaches the observer. The ubiquitousness of the shift to the red end of the spectrum, the redshift as astronomers call it, is due to the expansion of the universe, discovered by Edwin Hubble in the 1920s. Hubble's law states that the farther a galaxy lies away from us, the faster it is receding from us.

By the spring of 1999, Perlmutter's team had amassed a data set of 80 Type Ia supernovae occurring within galaxies that were much farther than those observed by Hubble and his successors. These were all exploding stars within galaxies whose light has traveled about seven billion years to reach us. In any one galaxy, with its billions of stars, a Type Ia supernova occurs only about once a century. How could the team have obtained 80 such images? The team's success was due to the cleverness of Perlmutter's search technique.

Even at such a low rate of occurrence, the laws of probability dictate that if we could look at a large enough set of galaxies, we should find some of these white dwarfs exploding at any given moment. So among tens of thousands of galaxies observed all at once, a few scores of supernovae were always detected. The three-week wait between the two successive observations of the same area of deep space had a purpose as well. A Type Ia supernova brightens up for about 18 days, and then fades away over the following month. Because of time dilation (a consequence of the special theory of relativity, as these galaxies recede from us at about half the speed of light), it appears to us on Earth that the supernovae undergo most of their brightening over a period of three weeks. Thus, observing the faraway galaxies within a three-week interval allowed the astronomers to "capture" and study the supernovae that had occurred during the time interval between the taking of the two electronic images.

But now, looking out of his window over the bay at the disappearing Sun and the fog rolling in through the Golden Gate, Perlmutter was distressed. Something made no sense to him at all. Ever since the big bang theory was proposed in the 1920s to explain the expansion of the universe, various theories have been advanced to explain what happened, how it happened, and to offer prescient ideas about the future of the universe. Einstein's equations predicted several scenarios.

First, the universe could be closed. In such a case, universal expansion eventually stops and the universe begins a collapse into itself because of the mutual gravitational attraction of all the matter in the universe. Second, the universe could slow down its expansion until it reaches a steady state and remain at that state. Scientists, and the public in general, seemed to favor the first scenario. Philosophically, there was something comforting in the belief that even though the Sun will die around five billion years from now, some day in the very, very distant future, the universe might start to collapse again and possibly—having finished one complete cycle of big bang birth and big crunch collapse—re-explode in a new big bang, which might create another Earth and a rebirth of life.

An expansion that slows down to a steady state was less favored by science, but was seen as possible. This would occur if the mass in the universe was just enough to halt the expansion, but not enough for the gravitational force to make everything fall back together again.

Only a few scientists believed in the viability of yet a third option—that the universal expansion would continue forever. And virtually no one imagined the unthinkable: that the rate of expansion of the universe would actually accelerate. And yet, mulling this unexpected possibility, Perlmutter could not ignore what his data were telling him. The faraway supernovae—along with their home galaxies—were receding from Earth at rates that were *slower* than expected. These rates were slower than the rate of recession of the more *nearby* galaxies. This could mean only one thing, he concluded: the universe was accelerating its expansion.

The reason for this baffling finding is not obvious. It has to do with the concept of *time*. The following is a simplified explanation, which leaves out some details. When an astronomer observes a galaxy seven billion light years away, he or she is seeing the galaxy as it was when the light left it to travel toward us, seven billion years ago. Thus the speed one computes for the galaxy from its observed redshift is the speed at which the galaxy was darting away from us *seven billion years ago*. Similarly, the recession speed for a galaxy one billion light years away is the speed of expansion a billion years ago. Now if the faraway galaxy is moving away from us at a slower speed than a closer galaxy, then seven billion years ago the speed of recession—the rate of expansion of the universe—was slower than the rate of expansion of the universe one billion years ago.[1] In other words, the universe is accelerating its expansion.

Perlmutter was befuddled. He had started this whole project years earlier hoping to measure the rate of *deceleration* of the universe—he never really expected our universe to be expanding *faster* all the time. There was something fundamentally disturbing about such a conclusion. And this was when Perlmutter started thinking about the sunset he was watching. Sunsets are red, the sky is blue—Rayleigh's Blue Sky Law that every beginning physics student recites. The atmosphere absorbs the white spectrum of light at varying degrees, depending on the light's frequency. Red light, with its low frequency and long wavelength, passes through dust and air particles more easily than blue light. Perlmutter is a careful scientist, and every scientist must watch out for possible errors in data. This is especially true for a scientist about to make a monumental conclusion about the universe—arguably the most important discovery in astronomy since Hubble's findings almost seventy years earlier.

What Saul Perlmutter found puzzling here was the apparently exceptional quality of his data. He had half-expected his data to be corrupted by the usual observational errors. The distant

1. The actual computation is more complicated. We observe light that left its galaxy seven billion years ago. When the light left its source, the galaxy that emitted it was actually about five billion light years from us. When its light arrives here, that same galaxy is at a distance of about twelve billion light years from us. The reason for the discrepancies is that space continues to expand all the time. Mathematically, the redshift we observe is a function of the entire stretching of space that has taken place from the moment the light left its source until the moment we received it (the redshift does not depend only on the instantaneous speed of recession of the light source).

galaxies his team had been observing should have had some dust in them, similar to the dust we find in our own Milky Way. The dust particles would have made the ancient exploding stars the team had been observing appear red as the sunset—and yet the supernovae were bright throughout the entire visible light spectrum (not counting the redshift—which shifts all the lines in a star's spectrum uniformly). This phenomenon told Perlmutter that there was little or no dust between the observers on Earth and their exploding stars halfway back to the big bang, and thus the observations were of exceptionally high quality. What the data were telling him had to be believed—the universe was expanding faster and faster. And this meant something frightening: our universe is infinite.

"Imagine a lattice in three dimensions," Perlmutter said to me shortly after he announced his group's extraordinary findings. "At each corner there is a galaxy. Now imagine that the lattice is growing. The distances from our corner, our galaxy, to all the other corners of the lattice keep increasing." The rate of increase—the rate at which space is being created between every corner and its neighbors—is accelerating. Since space is being created faster and faster, there is nothing to stop it, and it will therefore continue to expand forever. Our universe will expand to infinity. In a billion years, the distances from us to the faraway galaxies will be much larger and a billion years later it will grow even more than it did in the first billion years, and so on and on forever.[2]

2. There are local exceptions to this rule. Neighboring galaxies may have paths that bring them together despite the global expansion of the entire universe. Thus the Andromeda galaxy, which at a distance of 2.2 million light years from us is our closest neighbor (with the exception of the Large and Small Magellanic Clouds, considered offshoots of the

The data seemed to be clean of errors, and the message was clear. It was time to announce the news to humanity. This was done at the January, 1998, meeting of the American Astronomical Society.[3] The world was stunned. An infinite, ever-faster expanding universe was not what people had expected. Even many scientists had secretly hoped for a self-renewing universe, alternating eons of expansion and collapse—a cosmic garden going through the seasons. Instead, it seemed, the universe was fated to expand and fade away.[4] Stars would live out their lives and explode in supernovae or shed off their atmospheres as planetary nebulae. In our galaxy, new stars are born from the remains of dead ones. And the richness of the chemical elements produced inside dying stars is what allowed life to develop. But if the expansion were to continue and the density of space decline, eventually, after trillions of years, the universe would be a stellar graveyard of neutron stars and black holes.

What puzzled scientists was the question: Why? What was the explanation for the unprecedented new findings? The answer seemed to be that there is yet another mysterious force in the universe—something which had never been directly observed. That something, which physicists call a negative pressure or a

---

Milky Way), is moving in a path relative to the Milky Way which will make it collide with our galaxy in a billion years or so. Astronomers have found "rivers" of other galaxies that move in a direction counter to the general expansion of the universe.

3. A rival team of astronomers, which had used Perlmutter's ingenious methods to analyze its own smaller data set, announced similar results two months later.

4. General relativity does not imply that a big bang will follow a big crunch. Quantum effects, however, might make this possible if a universe re-collapses.

vacuum energy, or just a "funny energy," was counteracting the attracting force of gravity. There was something out there that was pushing away the galaxies—accelerating them on their mutual retreat from each other.

At the January 1998 meeting where Perlmutter announced his team's astounding results, other scientists presented findings—based on different methods of analysis—with the same puzzling inference. Astronomers Neta Bahcall and Xiaohui Fan of Princeton University, who studied massive galaxy clusters several billion light years from Earth, announced results that could also imply an ever-expanding universe. Based on studies of the mass density of galaxy clusters using three different techniques, Neta Bahcall and her colleagues found that we live in a lightweight universe. All of their studies independently showed that the universe probably has only about 20% of the mass density that would be needed to effect a collapse toward a new big bang.

Erick Guerra and Ruth Daly, also of Princeton University, obtained similar results from the study of 14 radio galaxies. Their analysis indicated once again that the mass of the universe is probably smaller than would be needed to stop the expansion some day in the distant future. All of these findings presented at the meeting brought back to life a haunting scientific concept, long ago placed in the dustbin of history.

<>

Cosmologists and astronomers arranged an urgent meeting to discuss the new results at the Fermilab near Chicago. The meeting on May 4, 1998, was organized by Paul Steinhardt, a bright young cosmologist now at Princeton University. Perplexed scientists from around the world converged on Chicago to discuss the reported acceleration of the expansion of the universe, and the fact that the universe may contain too little mass. Could equations be fit to the new data to explain them? Einstein's field equation for gravitation was the natural tool for the scientists to use for this purpose. But it would not explain the accelerating expansion—unless an old term, long ago discarded by the equation's discoverer and thereafter known as "Einstein's greatest blunder," was added back into Einstein's equation. The cosmological constant was back.

CHAPTER 2

Early Einstein

"Raffiniert ist der Herr Gott, aber boshaft ist er nicht."[1]

—Albert Einstein

The cosmological constant was an element that Einstein inserted into—and later removed from—his field equation of gravitation. This field equation was the crown jewel of Einstein's work, the culmination of the general theory of relativity he developed in the second decade of the twentieth century. It was an equation of such power, such insight into the latent laws of nature no one could see before Einstein, that its prescient nature appears staggering. Every decade since its inception, the equation has revealed its truth again and again in unex-

1. This is one of the most often quoted—and most badly translated—sayings in the history of science. The German statement was made by Einstein in 1921 on his first visit to the United States upon hearing a rumor, later proven false, that a non-zero ether drift had been discovered. Such a drift would have jeopardized the validity of the entire theory of special relativity. Translated literally, the statement says: "Tricky (crafty, shrewd) is the Lord God, but malicious He is not." However, the quotation is normally translated as: "Subtle is the Lord, but malicious He is not." I believe that the correct translation tells us more about Einstein's personal relationship with his God.

pected ways. How did one man understand the secrets of our universe so well?

Albert Einstein (1879–1955) was born in Ulm, a city in the region of Swabia in southwest Germany, on March 14, 1879, to middle-class Jewish parents whose ancestors had lived in the area for as many generations as anyone could remember. When Albert was still a baby, the family moved to the big city of southern Germany, Münich. There, his father, Hermann Einstein (1847–1902), owned a small business together with his brother, who lived with the family. The two were in the electrochemical industry and ran a small factory, Hermann doing the business side of things while his brother was taking care of technical engineering problems. Albert's mother was Pauline (Koch) Einstein (1858–1920). The Einsteins also had a younger daughter, Maja.

From an early age, young Albert showed great interest in the world around him. When he was five years old, his father gave him a compass, and the child became enchanted by the device and intrigued by the fact that the needle followed an invisible field to point always in the direction of the north pole. Reminiscing in old age, Einstein mentioned this incident as one of the factors that perhaps motivated him years later to study the gravitational field. From the ages of six through thirteen, Albert took violin lessons, encouraged by his mother, who had musical talents. Einstein became a good violinist and continued to play the instrument throughout his life. In 1886–1888, Albert Einstein attended the public school in Munich. Due to a state legal requirement about religion, his family had to supplement his education with Jewish education in the home, even though the family was not very observant. In 1888, Einstein entered the Luitpold Gymnasium in Munich, whose building was destroyed by bombing in the Second World War; it was rebuilt at another location and renamed the Albert Einstein Gymnasium.

At the gymnasium Einstein developed his great dislike for and distrust of authority—a trait he would exhibit throughout his life. Speaking later of his school years, Einstein compared the teachers of the elementary school to army sergeants and those of the gymnasium to lieutenants. It was this distaste for oppressive Prussian authority that made the young Albert renounce his German citizenship a few years later and apply for a Swiss citizenship. He would recall that the gymnasium used methods of fear and force and authority. It was here at the gymnasium that Einstein taught himself to question authority—and, in fact, to question all accepted beliefs, a notion that may have affected his scientific development as well, as some biographers have speculated. In 1891, another defining event occurred which, like the excitement over the compass, deeply influenced Einstein. As the textbook for one of his classes, Einstein was assigned a book on Euclidean geometry. He got the book before school began and read it through with amazement. Einstein had begun to question the premises of Euclidean geometry. Within two decades, he developed a revolutionary theory based on the view that the space in which we live is non-Euclidean.

In 1894, Einstein's family moved to Italy. His father had hoped to build a successful business there after the one in Munich had failed. The parents took Maja with them, but left Albert to stay with a distant relative so he could finish the gymnasium. Einstein decided on his own, however, to leave the school and rejoin his family in Italy. He couldn't stand the harsh, arbitrary discipline of the gymnasium, and he was bored studying the subjects that were emphasized: classical Greek and Latin. He longed to study more mathematics and physics—subjects in which he had been interested since early childhood. After six months, Einstein devised an escape. A doctor gave him a letter saying that he needed time off to join his family in Italy because of a breakdown. It seemed that the school was relieved to let him go since his behavior was disrupting the institutional discipline.

Einstein liked Italy very much. The prevailing culture—a celebration of things that made life worth living—stood in sharp contrast to the Teutonic order he despised. And he was enchanted with the wonderful works of art he saw in the museums all over northern Italy. Einstein even hiked over the Apennines from Milan, where his family lived, all the way to Genoa on the Ligurian coast of the Mediterranean. But Hermann's business failed once again, and he had to push his son to return to reality and to obtain a school diploma that would allow him to continue his education and support himself. The young Albert believed that his excellent knowledge of mathematics and physics should allow him to be accepted at a university without a diploma from the gymnasium he had deserted. But here he was wrong—it became clear that he would not be able to go anywhere without a diploma.

In 1895, Einstein failed the entrance examinations to the Swiss Federal Institute of Technology (known by its German acronym, ETH), where he had hoped to enroll by passing an exam in lieu of a diploma. He had done exceptionally well in mathematics, but his knowledge in other areas, such as languages, botany, and zoology, was below the institute's standards. Nonetheless, the director of the institute, impressed with the young man's knowledge in mathematics, suggested that he enroll at the cantonal school in the town of Aarau, Switzerland, to obtain the necessary diploma. Einstein enrolled at the school in Aarau with trepidation—he was still traumatized by the incomprehensibly mechanistic treatment the students received at the German gymnasium. To his surprise, however, he found the Swiss school to be completely different. Military-like discipline was not stressed here as it was in the German school. Here he was able to relax, study well, and make friends. He lived at the house of one of his teachers, and was a good friend to the teacher's son and daughter, with whom he made trips to the mountains. After a year at the Aarau school, Einstein obtained the diploma and applied to the ETH, where he was now accepted. He decided to study mathematics and physics and prepare for a career as a teacher. He was fascinated with the idea of explaining the natural world with concise mathematical expressions. To him, physics was a science aimed at finding an elegant mathematical equation to capture reality.

On October 29, 1896, Einstein moved to Zürich and enrolled at the ETH. Here he would meet two important people in his life, both fellow students at the ETH: Mileva Maric—later to become his first wife—and Marcel Grossmann, a mathematician whose work would help Einstein develop his theory of relativity in the years following his graduation. In his second year at the ETH, Einstein also met Michele Angelo Besso, who became a lifelong friend and a sounding board for Einstein's first ideas in developing the special theory of relativity.

During his first year at the ETH Albert Einstein made a decisive change of direction in his scientific career. Until then, he had been interested in mathematics and was proud of his knowledge of the discipline. At the Polytechnic, however, he decided that he was mostly interested in physics, and that mathematics was simply the way of quantifying physical laws. It was a vehicle for writing down in a concise manner the laws of the universe discovered through the science of physics. But Einstein was not happy with the instruction at the ETH. The physics professors taught old theories and did not discuss new developments in the field. Einstein started to do what he would continue doing throughout his life—teach himself theories by independent reading and study. Consequently he did not pay close attention to the lectures, and managed to antagonize many of the faculty. In mathematics, the situation was even worse. Having decided that mathematics was to be a vehicle and not a discipline of interest in its own right, Einstein paid very little attention to the instruction. This effect was most pronounced during lectures by Hermann Minkowski (1909–1964), a renowned mathematician of Russian origin. Minkowski was so put off by the young student's cavalier attitude to his classes that he later described him as "a lazy dog." As fate would have it, when Einstein developed his special theory of relativity within a few years of graduating from the ETH, it was Minkowski who developed an entire area of mathematics to describe the physics of relativity.

Einstein's carelessness with his education at the ETH backfired upon graduation. As every undergraduate at a university today knows, attending classes and getting good grades are important, but there is something at least as important for one's career—the ability to obtain excellent letters of recommendation from faculty members. In Einstein's day, this need was even more acute. In order to continue graduate studies at a prestigious institution, a student had to obtain the recommendation of a faculty member with whom he or she had worked as an assistant. To Einstein's great disappointment, not a single one of his professors agreed to take him on. Einstein had to leave the ETH and find a position as teacher or tutor. His situation was more dire due to his father's financial difficulties and his family's inability to give him any support even while he was looking for a job.

Einstein graduated from the ETH in the summer of 1900, but since he failed to obtain a position as an assistant at the institute, he had to look hard for some means of support. During the next few years, he held temporary teaching positions in Switzerland, but never anything long-term.

On June 16, 1902, Albert Einstein, now a Swiss citizen for over a year, was hired to work at the Swiss patent office at Bern, a position arranged by the father of his good friend Marcel Grossmann. The position was at first temporary, but in 1904 it became permanent. He was named a Technical Expert, and his job was to evaluate the merits of patent proposals. The previous two years had seen changes in Albert's life: the death of his father in Milan in 1902, and Albert's marriage to Mileva in 1903. Mileva had followed him to Bern, and they married despite Einstein's mother's objections and her dislike for his fiancée.

The Swiss patent office presented an interesting opportunity for the young scientist. He seems to have enjoyed his work. Throughout his life, Einstein reveled in tinkering with devices invented for particular purposes and trying to evaluate their usefulness. The position left him some free time—time he used well for study and research. Later in life, he suggested to young researchers that the best situation for a creative scientist is to have a menial or "unintellectual" job that allows some free time for research, rather than a traditional university position requiring teaching, service to the institution, and campus politics.

Einstein spent much of his time at the Swiss patent office reading and doing research. Contrary to what some of his biographers have said, Einstein was well aware of the work of his contemporaries as well as earlier physicists and other scientists, and he also read the work of well-known philosophers, including Immanuel Kant, Auguste Comte, David Hume, and Nietzsche. In physics, Einstein was most affected by the works of Galileo Galilei (1564–1642), Ernst Mach (1838–1916) and James Clerk Maxwell (1831–1879). Galileo was the first to consider the relativity of moving systems, and in his developing work Einstein would often refer to Galilean reference frames. The Austrian physicist Ernst Mach conducted thorough analyses of the mechanics of Isaac Newton (1643–1727). Mach noted that Newton organized observations of movements under several simple principles from which he then carried out predictions. But Mach asked whether these predictions were only correct as long as the experiences Newton described were correct. Mach emphasized that in science we must follow an economy of thought—build models that are parsimonious, having as few parameters as possible. This is the mathematical form of Occam's Razor, the well-known principle that the simplest theory has the best chance of being correct. In mathematical sciences, this means that one should choose the simplest model, or equation, to describe any phenomenon in nature. In a way that anticipated Einstein's major work, Mach criticized Newton's reliance on the concepts of absolute space and absolute time. In this way, Mach's philosophy of science is relativistic, although he was an early opponent of atomic theory, since there were no direct observations that proved its existence at the time (the 1870s). For Mach, all scientific conclusions were distilled from physical observations. Drawing on Mach's insistence on the empirical, Einstein made physics relative and precise, and left Newton's theory as the limit of relativity when speeds are those encountered in everyday life.

The scientist who influenced Einstein's work the most was the Scottish physicist James Clerk Maxwell. Maxwell developed the idea of a *field*, which is essential to all of Albert Einstein's work. Maxwell's theory explained electromagnetic phenomena by a system of equations describing a field of forces—like the lines one sees when a magnet is placed under a piece of paper on which iron shavings have been spread. The iron shavings align themselves in a distinct pattern from one magnetic pole to the other. The visible patterns are a picture of the magnetic field produced by a magnet. Maxwell's work opened the way for science to do away with fictitious concepts such as the ether, which was believed to be the invisible medium through which light moved in space. Maxwell's work can be seen as the harbinger of Einstein's relativity, where fields are basic elements of the theory. But other scientists contributed as well to Einstein's fund of knowledge, which he used in developing the special theory of relativity while employed at the Swiss patent office. These scientists included Heinrich Herz (1857–1894); the Dutch physicist Hendrik Lorentz (1853–1928), whose work on transformations was crucial in explaining the mathematics of the special theory of relativity; the great French mathematician Henri Poincaré (1854–1912); and several others.

Einstein put forward the special theory of relativity in 1905, a year in which he also completed three other groundbreaking works which were published that amazing year: papers on Brownian motion, on the light-quantum theory, and a doctoral dissertation on molecular dimensions. Einstein's work on relativity changed our notions of motion, space, and time. Space was never again to be viewed as absolute, but rather relative to one's frame of reference. The idea of reference frames echoed a similar notion proposed by Galileo three centuries earlier. Galileo considered what would happen if a ball was dropped from the top of a mast of a ship, as compared with a ball dropped from the same height on land. In the first case, the reference frame—the boat—is moving, while in the second case the reference frame—terra firma—is not moving. What would happen to the ball? asked Galileo. Would it fall straight down on the moving boat, or trace a path backwards as if it had been dropped on solid ground? Einstein took this idea of moving frames of reference and brought it to unexplored territory—objects moving at speeds close to that of light.

In this new relativistic world Einstein had given us there was only one absolute: the speed of light. Everything else was wrapped around that ultimate speed limit. Space and time were united to give us spacetime. A twin traveling on a fast spaceship was proved to age more slowly than his or her twin remaining on the ground. Moving objects change and time dilates as the speed of a body approaches that of light. Time slows down. If anything could go faster than light, which relativity forbids, it would then move into the past. Space and time are no longer rigid—they are plastic and depend on how close an object gets to the speed of light.

The absoluteness and universality of time had been a sacred tenet of physics and no one had questioned the assumptions. Time was the same everywhere, and the flow of time was constant. Einstein showed that these assumptions were simply not true. The constant quantity was the speed of light—everything else, space and time, adjusted themselves around this universal constant. Einstein's special theory of relativity made obvious one of the most puzzling negative experimental results in history: the Michelson-Morley quest for the ether.

James Clerk Maxwell, who had done so much for our understanding of physics, and whose theory inspired Einstein, was-like other scientists in the pre-relativity world—a believer in the theory of the ether, derived from the ancient Greeks. In an entry he wrote for the 1878 edition of the *Encyclopedia Britannica*, Maxwell wrote: "All space has been filled three or four times over with aethers." But what is this aether, or ether? People thought that light and other radiation and particles needed some medium through which to travel. No such medium was ever actually seen or felt, but somehow, it had to exist. This assumption was so pervasive that respected scientists took it very seriously. One of them was Albert A. Michelson (1852–1931), a noted American physicist who in 1881 was working at a Berlin laboratory and was made aware of a letter written by Maxwell

Albert Einstein,
November 1921
AIP Emilio Segrè Archives, Segrè Collection

in 1879, in which Maxwell inquired whether astronomical measurements could be used to detect the velocity of the solar system through the ether. Michelson was an expert at measuring the speed of light. He was intrigued. Michelson embarked on a series of work of increasing accuracy aimed at detecting changes in the speed of light which would indicate an ether drift. Most of the experiment was done jointly with the American chemist Edward W. Morley (1838–1923) in 1886, after Michelson's return to the United States. Michelson and Morley measured the speed of light both in the direction of the Earth's rotation and against that direction and expected to find a difference in the speed. But they did not. There was no ether drift, and apparently no ether. In 1907, Michelson became the first American scientist to receive a Nobel Prize. By then, Einstein's special theory of relativity had explained to the world *why* Michelson and Morley had obtained their unexpected result.

It is not clear when Einstein learned of the Michelson-Morley experiment that found the surprising zero change in the speed of light when measured with or without Earth's rotation. Einstein used purely theoretical considerations, his "thought experiments," to determine that the speed of light remains constant no matter how fast the source of light moves toward or away from the observer. Einstein's biographer Albrecht Fölsing describes the day in mid-May 1905 when the final realization of the principle of special relativity came to Einstein at the patent office in Bern.[2] It was a beautiful day, Einstein later recalled in a lecture at Kyoto in 1922. He had spent many hours discussing the problem of space and time with his friend Michele Angelo Besso, and then, suddenly, the answer came to him. The following day, without saying hello, Einstein pounced on his friend with the explanation of the principle of relativity: "Thank you! I've completely solved the problem. An analysis of the concept of time was my solution. Time cannot be absolutely defined, and there is an inseparable relation between time and signal velocity." Einstein explained to Besso the idea of simultaneity. Within relativity, time is not the same everywhere. Einstein used the bell tower of Bern and the bell tower of a neighboring village to illustrate his point. The constant was not time, not even space—it was the speed of light. And the special theory of relativity explained it all. But what if an ether drift were detected? It was

2. A. Fölsing, *Albert Einstein*, New York: Penguin, 1997, p. 155.

upon hearing a rumor of such an experiment years later, in 1921 when relativity was accepted by most of the world, that Einstein made his now-famous statement "Raffiniert ist der Herr Gott, aber boshaft ist er nicht." The saying is chiseled in stone over a fireplace in the common room of the mathematics department at Princeton University—a testimony to the enduring nature of the special theory of relativity.

CHAPTER 3

Prague, 1911

"If Einstein's theory should prove to be correct, as I expect it will, he will be considered the Copernicus of the twentieth century."
—Max Planck[1]

Einstein realized that relativity—the "special" theory of relativity he had constructed—was true in a world without massive objects. Masses, and gravitation, required another theory. The existing theory of gravitation was the one developed by Isaac Newton three centuries earlier, but once the special theory of relativity was understood, it was clear that Newton's theory was only a limiting case, correct for a world where speeds are much less than that of light. So there were two theories, Einstein concluded, special relativity and Newton's gravitation. Both of them were good in special limiting cases: Newton's theory was good in a low-speed world but would have to be corrected in a universe where light and its speed—the universal limit—play a role. Similarly, special relativity was correct when gravity was insignificant, and that theory, too, would have to change so it would be true in a universe dominated by massive objects. If the speed of light is absolute, and *time* itself is relative, then Newton's laws could not hold under such con-

1. In a letter to the faculty committee at Prague, 1910.

ditions when the special theory of relativity was relevant, that is, when speeds approach that of light. In such cases, where time becomes relative, the rules for moving objects could not be the old Newtonian laws, Einstein deduced. Somehow, the two theories—Newton's theory of gravitation and Einstein's special relativity—had to merge, to give a *general* theory of relativity. This would be a theory of relativity *and* gravitation. But how could this be achieved?

In 1907, having derived the principle of special relativity two years earlier, Albert Einstein, now 28 and working at the Swiss patent office in Bern as a Technical Expert, second class (promoted from third class just the year before), now directed his attention to the problem of gravitation.

Some time in November 1907, Albert Einstein was sitting in his chair at the patent office in Bern, thinking about the implications of the special theory of relativity, whose ideas he had finished developing two years earlier. He later described the portentous moment in his 1922 Kyoto lecture with these words: "All of a sudden, a thought occurred to me: if a person falls freely, he will not feel his own weight. I was startled. This simple thought made a deep impression on me. It impelled me toward a theory of gravitation." To his close friend, Michele Angelo Besso, who also worked at the Swiss patent office, Einstein described the revelation as "The happiest thought of my life." Einstein sought to explain the force of gravity within the theory of relativity. Ultimately, this would lead him to the development of the general theory of relativity—a theory of relativity that would incorporate gravitation.

<>

For four years, from 1907 until June 1911, Einstein remained mysteriously silent on the issue of gravitation. In 1911, he moved from Switzerland to Prague. It is not known whether Einstein worked on the problem of gravitation during the four intervening years. During these years, he published papers on blackbody radiation, and on critical opalescence, but was the important problem of gravitation and its relation to relativity on his mind? The discoverer of the special theory of relativity, the theory that revolutionized the way physicists viewed the universe, and the man who by then had contributed significantly to all areas of physics, was still on insecure professional grounds. Einstein's salary was always modest, and he even tried to supplement it by teaching at the university in Bern, something he viewed as an unpleasant burden. Later in life he lamented that he, who in his mind placed many clocks at various places in space and imagined them flying at different speeds, deducing that space and time were relative, did not have enough money to buy a single clock for his home.

On April 4, 1910, Einstein wrote a cryptic letter to his mother from Zürich, where he was an associate professor. "I will most probably receive a call from a large university to be a full professor with a salary significantly better than I now have. I am not yet permitted to say where it is."[2] Einstein told some of his colleagues the same thing, and later that year the mystery was revealed: the university was the German University of Prague. Thus Einstein, who as a teenager had renounced his German citizenship to become a Swiss, and who had developed the special theory of relativity entirely on Swiss soil, now prepared to join a German university, and to begin a track that would ultimately lead him back to Germany, to the capital of the German state he deplored, Berlin.

2. Letter reprinted in A. Pais, *Subtle Is the Lord,* New York: Oxford University Press, 1982.

The German University of Prague had an unusual history, and one that reflected the sad state of inter-ethnic relations in the Bohemian capital at that time. The university was the oldest one in eastern Europe, and in the nineteenth century it employed both Czech and German professors. The two groups, however, never got along with each other—to the point that the German faculty did not even exchange professional information with its Czech counterpart. In 1888, the Austro-Hungarian emperor decreed that the university should split into two halves: a German and a Czech half. The split caused a further rift between the two faculties and contributed to more hostility between the two groups. Einstein was hired by the German university.

Prague, an important city of the Hapsburg empire, and a corner of the Vienna-Budapest-Prague triangle from which the emperor ruled the Austro-Hungarian realm, held an attraction for Einstein. So much so, in fact, that he stood firm in his decision to relocate there despite the fact that he knew he would be isolated from the centers of scientific research, and despite Zürich's offer to raise his pay to compete with the wages promised him by the Karl-Ferdinand University of Prague. Einstein was undeterred by manifestations of anti-Semitism, evidenced from committee discussions of his ethnic origin, which came to his attention during his application process. He was made to state his religion on his application for the professorship, and the university would not accept the "none," he had originally put down as answer. Emperor Franz Josef himself signed the decrees appointing faculty to their posts, and it was known that the emperor would not appoint anyone whose religion wasn't stated on the application form. Einstein relented, the emperor decreed, and the appointment became effective April 1. Possibly being pushed on a matter of religion made Einstein, who'd never before exhibited religious feelings, join the

Prague Jewish Community. He is also known to have visited the famous Old Jewish Cemetery of Prague, whose first use dates back to the fifth century, and to have examined the crumbling stone above the grave of Rabbi Löwe, friend of the sixteenth century astronomer Tycho Brahe.

It seems from his letters to friends that Einstein was not happy in Prague. He often complained about the bureaucratic red tape and the Prussian rigidity of the German officials who ran the university. He also felt that the students were not nearly as intelligent and industrious as the ones he had taught in Switzerland. Whether or not he liked Prague, Einstein seems to have left his mark even on Prague's social life. In his book *Prague in Black and Gold*, Peter Demetz writes about the café where Einstein liked to spend his free time.[3] Demetz writes that even café life in Prague was segmented into Czech- and German-patronized establishments. The Café Slavia was Prague's showcase, a favorite haunt of famous Czech linguists as well as writers, among them Thomas Mann. Liberal journalists usually crowded in the back corner, and progressive Catholics in the front by the sidewalk. It was at this trendy establishment that Einstein would often be seen on sunny afternoons, talking German with colleagues from the university or copiously filling pieces of paper with equations. It was here in Prague, city of Kafka, of café life, and of perpetual intrigue between the Austro-Hungarian administration and the Jesuits, that Einstein made important first steps toward his general theory of relativity.

The first concept Einstein addressed in Prague was the equivalence principle he had first stated in Bern four years earlier.

3. Peter Demetz, *Prague in Black and Gold*, New York: Hill and Wang, 1997, p. 354.

Einstein imagined two frames of reference: one at rest carrying a gravitational field, the other a field-free frame in constant acceleration. In both frames, Newton's laws must be the same, and the equivalence should be derived from a new theory of gravitation, Einstein stated in a paper published that year. Therefore, his aim was to look for a new theory—one that would encompass both notions of gravitation and relativity.

The next principle Einstein derived in Prague was the redshift of gravity. He started from the equivalence principle and deduced that a ray of light emanating from a massive object will drop in its frequency and shift toward the red end of the spectrum. In 1911, Einstein knew that special relativity, where a redshift of light occurred due to speed of the source away from the point of observation, should be incorporated into the theory of gravitation. But he didn't know how. At that point, he lacked the tools to derive how gravitation, too, causes the redshift of radiation. He was, however, able to demonstrate that a redshift due to gravitation must exist.

When Einstein discovered the special principle of relativity, the mathematics was there for him: the Lorentz transformation, and the mathematics of space-time worked out by Minkowski. Minkowski's mathematics entwined the three independent directions of space with the arrow of time. This design allowed for a uniform treatment of the four components of Einstein's space-time, where events and connections between present, past, and future are made through cones in four-dimensional space. A two-dimensional analogy is shown below. The light cone reflects the constancy of the speed of light starting at the origin. On the light cone, the space-distance from the origin is equal to the time elapsed. The hyperboloid in the figure is the set of all points with equal squared space-time distance from the origin. The Minkowski metric allows for measuring distances in spacetime.

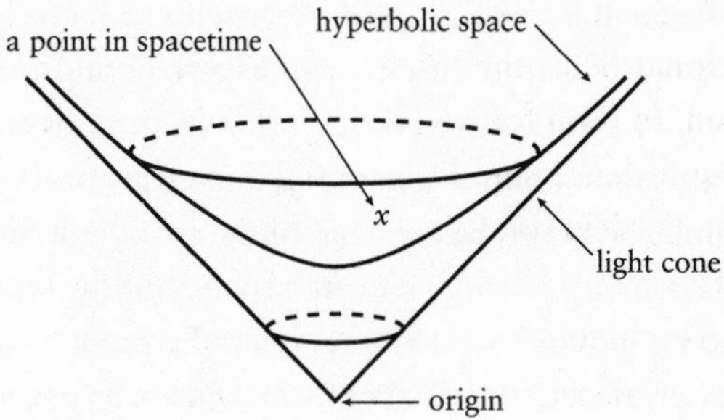

Minkowski space

The mathematical application was novel, but the mathematics itself was not very complicated, since its basic elements—called vectors—had long been well-understood. But now in Prague, as he returned to the idea of incorporating the concept of gravity within the special theory of relativity, Einstein realized that he needed powerful mathematics, much more powerful than the mathematics used in special relativity, and in areas about which he knew very little.

Gravity made space non-Euclidean, and therefore Einstein would need some new geometrical tools to handle this curvature. Einstein now had to master very complex mathematics. He began the hardest part of his journey of discovery, an endeavor that required him to muster every physical intuition and wed it to the powerful machinery of mathematics.

The needed tool for Einstein's next step in the long journey was hiding in plain sight right there in Prague, in the person of a gifted yet perennially underappreciated mathematician. Georg Pick was twenty years older than Einstein, and the two of them had met shortly after Einstein joined the faculty in Prague. The custom in Prague was that every new professor to join the university call on all other members of the faculty at their homes. Since Einstein was already well-known in academic circles, the forty-odd members of the faculty waited expectantly for his calls.[4] As a newcomer to the city with its fascinating architecture and history, Einstein enjoyed the first few visits to his colleagues, since each visit took him to a new and interesting part of the city. After a while, however, he got tired of the small talk, which was keeping him away from general relativity, and he stopped the visits. Going down the list of names, Einstein presumably stopped after the letter "P," since Georg Pick was not one of the professors slighted by Einstein. Pick and Einstein became good friends and took long walks together, discussing mathematics. Pick was a source of stories about Ernst Mach, who had been at the university before Einstein's arrival, and whose ideas anticipated the emergence of Einstein's theory of special relativity. Pick, like Einstein, was a good violinist, and through him Einstein joined a local quartet. But Pick was also an expert on a mathematical method that Einstein needed in order to develop his general theory of relativity. Pick was familiar with the work of the two Italian mathematicians, Gregorio Ricci (1853–1925) and Tullio Levi-Civita (1873–1941). As early as 1911, Pick may have tried to direct Einstein to the mathematics of Ricci and Levi-Civita, but Einstein remained oblivious to the good advice and would remain so until after he left Prague. Had he looked at the mathematical papers of the two Italians, Einstein might have saved himself a few years of hard work.

The third piece of work on a general theory of gravitation that Einstein did in Prague was on the principle that massive objects affect not only rigid bodies, but also light. Einstein devel-

4. Philipp Frank, Einstein's best contemporary biographer, writes at length about this story and other curiosities of Einstein's life in *Einstein: His Life and Times,* New York: Knopf, 1957.

oped his first ideas here on a principle that would turn out to be equivalent to the one Newton had used centuries earlier. His deduction that light must bend around a massive object was tantamount to the Newtonian principle that an object, flying in space, would change its path when nearing a massive object. This is the same principle that allows NASA to change a spacecraft's direction by making it swing around a planet. The theory gave the amount of deflection that a light ray would experience when it passed by a massive object, assuming light was not a ray but a particle. For an object with the mass of the sun, and a ray of light just grazing its edge, Einstein computed a deflection of 0.83 arcsecond (a measure of angular separation). It appears that Einstein made an arithmetical error, for the value he should have gotten from the calculations is 0.875 arcsecond. This latter value is still only half of the correct value for the deflection, which Einstein would obtain four years later within a completed general theory of relativity.

Having made some progress and derived some principles—even if he had not completed work for a new theory—Einstein felt strongly that he needed physical proof of his theoretical findings. The deflection of light was something he was aware of while still in Switzerland, but then he was convinced that the effect was too small to ever be detected. He told other scientists of his conviction that light should be affected by gravity but that there was probably no way of ascertaining this effect experimentally. In Prague, Einstein had second thoughts about this problem. Once he had an actual number in his hand (alas, a wrong one, but still a positive deflection of a light ray), he wondered if astronomers could somehow measure this effect. He was interested in obtaining proof of the predictions of his nascent theory of gravitation. If the bending of light could be detected, that would provide a welcome proof of his theory.

Unbeknownst to Einstein, in 1801 a German astronomer, Johann Georg von Soldner, had the same idea. He intended to apply Newton's theory of gravitation to light rays as if they were massive objects. Soldner used the Newtonian scattering theory, which assumes that light consists of small particles. Soldner found, for the same problem addressed by Einstein a century later, that light passing close to the surface of the sun should be deflected by 0.84 arcsecond. This was amazingly close to Einstein's number, which was affected by an error. Soldner's deviation from the true Newtonian value of 0.875 arcsecond is probably due to an inaccurate estimate of the mass of the Sun. Soldner's work did not come to the attention of the physics community until 1921.

In his paper on the deflection of light, Einstein suggested that evidence for the phenomenon could be sought by astronomers. In the summer of 1911, a student at the Karl Ferdinand University in Prague, Leo W. Pollak, traveled to Berlin and visited the Berlin Observatory. There, he met Erwin Finlay Freundlich (1885–1964), who was the youngest assistant at the observatory. Freundlich was born in 1885 in Biebrich, Germany, of a German father and a Scottish mother. After earning a doctorate from Göttingen University, Freundlich took a position at the Berlin Observatory. Pollak mentioned to Freundlich when he met him that Einstein was disappointed that astronomers had not taken up his suggestion that the deflection of light could be detected experimentally. Pollak's description of the ideas in Einstein's paper caught Freundlich's interest and he offered to help.

Shortly after Pollak's visit, Freundlich wrote to Einstein in Prague and offered to conduct measurements of starlight passing by the planet Jupiter, to see if it was deflected by the planet's gravitation. The attempts failed, and on September 1, Einstein wrote a letter to Freundlich thanking him for his continuing efforts and lamenting the fact that there was not a planet larger than Jupiter. Despite the failed experiments, the cooperation between the two continued for many years.

Einstein spent the week of April 15 to 22, 1912, at the Royal Observatory of Berlin, visiting Freundlich. In 1997, Jürgen Renn of the Max Planck Institute for the History of Science, in Berlin, reported in the journal *Science* the results of research he and his colleagues have conducted of a notebook Einstein kept during his visit, and which had been previously unknown.[5] Among notes on daily appointments, Einstein wrote the essence of a startling discovery he had just made: gravitational lensing. This effect takes place when the light from a distant star or galaxy reaches the observer by way of an intervening star or galaxy. The bending of light which Einstein knew to take place can happen in a symmetrical way, where the light rays bend all around the intervening heavenly body. This effect focuses the light, just as light focuses when it passes through a glass lens. Thus, the light from a distant star can be magnified by the "gravitational lens" provided by the star lying between the observer and the distant object, allowing the observer to see the distant star better. Today, astronomers use gravitational lensing to observe very faint, faraway galaxies whose light happens to pass by closer galaxies and get focused by them. Computers are then used to disentangle the distorted light from the gravitational lens. Einstein is known not to have thought much of his discovery of 1912, thinking that the effect he had discovered would never be observed.

In 1936, at the persistent prodding of a Czech amateur scientist, Rudi W. Mandl, Einstein submitted a paper describing

5. Renn, J., et al., *Science*, January 10, 1997.

his theoretical development of this effect to the journal *Science*, after Mandl had asked him whether such an effect would be possible. It is not known whether Einstein actually remembered that he had derived the theory twenty-four years earlier in a notebook he had left in Berlin. From notes Einstein left in 1936, it seems that he had derived the theory all over again. In a letter to the editor of *Science* in 1936, Einstein wrote: "Some time ago, R. W. Mandl paid me a visit and asked me to publish the results of a little calculation, which I had made at his request. This note complies with his wish." Then in a more personal letter to the editor, James Cattell, Einstein wrote: "Let me also thank you for your cooperation with the little publication, which Mr. Mandl squeezed out of me. It is of little value, but it makes the poor guy happy." In 1979, the gravitational lensing effect was first observed by astronomers, causing much excitement. Today the effect is both studied in its own right and is used as an important tool for deep-space astronomical observation.

<>

Soon after Einstein's arrival in Prague, he received an offer for a full professorship at the polytechnic in Zürich (ETH), where he had studied. Einstein loved his adoptive Switzerland and so not long after settling in Prague, his eventual departure within a year was a foregone conclusion. Perhaps the ephemeral nature of his stay there allowed Einstein to be bolder in science and to experiment. And the topic on which he had chosen to concentrate his efforts would require years of work to perfect. Philipp Frank, who came to the University of Prague just before Einstein left, tells a number of amusing anecdotes about Einstein's almost surreal stint in Prague. The office Einstein was assigned at the university overlooked what seemed a beautiful, manicured green park. Looking through his window, deep in thought about the problems of gravity, Einstein couldn't fail to notice that in the morning only women walked in the park, while in the afternoon it was only men. Perplexed by this observation, Einstein asked people what was going on in the park below. He was told that this was no park—these were the grounds of a mental institution. Later he would joke with colleagues about the people in the park, saying they were the mad people who would not occupy themselves with the quantum theory. (Einstein had a lifelong battle with the quantum theory. In reference to the quantum theory, with its probabilistic nature, Einstein would later make his famous statement: "I shall never believe that God plays dice with the world.")

The general theory of relativity—the new theory of gravitation—was not a problem Einstein could solve in one year in Prague. For completing this theory, he would require another five years—and a lot more mathematics than he knew at that time. Einstein did manage to discover two important truths, outcomes of the rudiments of his developing theory. One is the redshift that occurs when a light ray goes through a gravitational field. The energy of the light ray is decreased as the gravitational field of a star, for example, pulls on the ray. Since the speed of light is constant—the basic tenet of the special theory of relativity—what is affected by the gravitational pull of the star is the frequency of the light and its related wavelength. The frequency decreases (fewer peaks of the light-wave per unit of time) and correspondingly the wavelength increases. Since the longer wavelength lies in the direction of red light rather than blue, this lengthening of the wavelength is called a redshift. In Prague, Einstein discovered theoretically the phenomenon of gravitational redshift. The second discovery within the theory he was developing was that light must bend around massive objects. Since a massive object such as a star makes space around it curved, that is, non-Euclidean, a light ray passing near such a massive body should bend and follow the curvature of space (although the amount he derived for the bending was that of Newton's theory, and thus wrong by half). With these two discoveries in his pocket, and work in other areas of physics, Einstein was ready to return to Switzerland.

Professors at the university in Prague were issued a uniform. While not required to wear the uniform daily, it was to be worn when taking the oath of allegiance before assuming their duties and at any time when in an audience with the Austro-Hungarian emperor. Throughout his life, Einstein abhorred authority and shunned protocol and ceremonies. The uniform made him uncomfortable, but he eased the situation by joking that if he wore it outside on the streets, people would take him for a Brazilian admiral. Einstein was all too happy to get rid of his uniform. He gave it as a present to his successor at the university, Philipp Frank, upon his arrival. Frank, in turn, wore the uniform only once—when he swore allegiance to the Austro-Hungarian emperor. In 1917, Frank's wife made him give Einstein's uniform to an ex-army officer, a refugee from the Russian revolution, who was freezing on the streets of Prague since he had no money to buy a coat.

As Einstein continued to study the problem of gravitation and to try to cast it within the framework of special relativity, he came to a startling conclusion: space is not Euclidean. In a paper written toward the end of his stay in Prague—just before his decision to accept an offer from the ETH for a full professorship and to return to Switzerland, Einstein wrote a paper published the following year in the journal *Annalen der Physik*. His paper stated a revolutionary conclusion from his investigations of space and gravity: the laws of Euclidean geometry do not hold in a uniformly rotating system. By the special theory of relativity, a contraction of the circumference would occur, and space would be warped. Straight lines would not exist as such, and the ratio of the circumference of the circle to its diameter would no longer be pi. Since, by the equivalence principle he had derived in Bern, a uniformly rotating system induces a field that is equivalent to a gravitational field, Einstein came to a stunning conclusion: near a massive object, space is not Euclidean. But what does "Euclidean" or "non-Euclidean" mean?

CHAPTER 4

Euclid's Riddle

"There is no royal road to geometry."
—Euclid of Alexandria to Ptolemy I,
King of Egypt, 306 B.C.

Cape Perpetua rises a thousand feet above sea level on the rugged Oregon coast, the high surf of the Pacific Ocean crashing vigorously and with clockwork regularity onto the craggy inlets below. Jutting into the air above a deep blue ocean, Cape Perpetua is unique. For a person standing at the top of the promontory will clearly see that the earth is round. The vast ocean ahead of the observer is seen to gently curve downward in every direction the eye can see. And as a boat sails away, it seems to the observer to glide down the rounded surface of the earth for the longest time, and gradually disappear behind the giant blue ball.

Had the ancient Babylonians, Egyptians, or Greeks lived on the Oregon coast, perhaps the history of mathematics and the exact sciences would have been very different. But these peoples of antiquity did not inhabit the Pacific coast and had never seen the curvature of the space in which we live. The Babylonians, and their relatives the Assyrians, lived on the flat lands between the Tigris and Euphrates rivers of Babylon, and their world was flat. From the thousands of clay tablets they left behind, detailing every aspect of life in their society as early as

4000 B.C., we know that the Babylonians were adept at computing exact areas of fields. They knew how to divide the flat arable land they possessed using a right-angle so that they could use simple multiplication of sides to find the area of a plot of land. They also knew how to find the areas of right-triangular fields by dividing the area of the inscribing rectangle by two. The Babylonians and Assyrians were experts at these aspects of plane geometry. The Egyptians, too, were very advanced in the geometry needed for marking and dividing and computing the areas of plots of land. But they also lived in a flat land and never saw the need to understand a surface that is not flat. Even their pyramids were masterpieces of straight-line geometry—alas in three dimensions.

In the sixth century B.C., Pythagoras and his followers in the hamlet they established in Crotona in southern Italy developed abstract theorems based on the applied work of ancient Egypt and Babylonia. The Pythagorean theorem is thus an extension of the Babylonian mathematical interpretations of the physical world. The theorem states that a square field whose side is the hypotenuse of a right triangle is equal in area to the square fields on the other two sides. The Pythagorean theorem has important implications in geometry, since it can be used to define the shortest distance between two points in Euclidean space. A straight line in such a space is the shortest distance between two points. If we know the difference in the X-direction between the two points, and we know the difference between the two points also along the Y-direction, then the shortest distance between two such points on the plane is the square root of the sum of the squared differences in the X and Y directions. But the Pythagoreans went much further and discovered irrational numbers. They noticed that when each side of the triangle is equal to one, the hypotenuse is a curious number—the square root of two, which was *irrational*: it could not be written as a ratio of two integers. The discovery of new numbers which they could not comprehend and which did not seem to hold any meaning within the physical world led the Pythagoreans to areas of mathematics that would be greatly developed in our own time.

Mathematics continued its growth, and two centuries after Pythagoras, Euclid of Alexandria wrote the *Elements*—a book in thirteen volumes considered the greatest textbook ever written. The volumes of the *Elements* laid out an entire geometrical theory—one that would guide the study of mathematics through twenty-three centuries to our own time. *Euclidean geometry* is an attempted abstraction of the notions of physical space with the aim of using axioms, postulates, and theorems to explore the essential properties of the space the ancients thought was the only one.

Euclid defined the elements of his geometry as a point, a line, and a plane—now familiar notions to anyone who has taken a class on elementary geometry. Euclid then laid out five main postulates: 1. To draw a straight line between two points; 2. To continuously produce a straight line; 3. To describe a circle with any center and radius; 4. That all right angles are equal to one another; and 5. That if a straight line falling on two straight lines makes the interior angles on the same side less than two right angles, then if the two straight lines continue indefinitely, they will meet on the side where the sum of the interior angles is less than two right angles (180 degrees).

The propositions, or theorems, in Euclid's first book deal with the properties of straight lines and the areas of parallelograms, triangles, and squares. While Euclid made essential use of his first four postulates in the proofs of the propositions, the fifth postulate was not used in any of them. It was evident quite early on that these propositions would still be valid if the fifth postulate was deleted or replaced by another one compatible with the other four. Although the *Elements* became an immensely popular book, one which has influenced Western thought for two millennia, the tenuous nature of the mysterious fifth postulate raised persistent questions in the minds of mathematicians. Even in its description, the fifth postulate is an oddity: while the other four are terse and clear, the fifth is lengthy. To many, the fifth postulate seemed more like a theorem to be proven than a self-evident fact.

The fifth postulate has several equivalent restatements. One, called Playfair's axiom, says that only one parallel to a given line can be drawn through a given point. Another equivalent of the fifth postulate is that the sum of the three angles of a triangle is always equal to two right angles (that is, 180 degrees). It is this latter consequence of the fifth postulate that is easiest to analyze.

From the very first appearance of the *Elements*, skepticism was voiced by geometers about the fifth postulate as a necessary or even true element of the entire theory. The first to voice the most serious complaint against Euclid was a geometer from whom we learn much of the history of the work. He is the fifth century philosopher, mathematician, and historian Proclus (A.D. 410–485). From Proclus we know that Euclid lived during the reign of the first Roman sovereign of Egypt, Ptolemy I, and that the king himself wrote a book about Euclid's problematic fifth postulate—including a proof of the postulate based on the other four. This was the first attempt we know of from historical sources to prove that the fifth postulate is a consequence of Euclid's four prior ones.

Proclus argues correctly in his history of Euclid's work that Ptolemy's very "proof" that postulate five is a consequence of the others makes use of the assumption that through a point not on a line there can be only one line parallel to the given line—an equivalent statement of postulate five! Then he attempts to give his own "proof" of the redundancy of the postulate. His proof was false as well.

Arab science flourished in the Middle Ages, after the great civilization of ancient Greece was no more, and before Europe woke up from the darkness of centuries. Omar Khayyam (ca. 1050–1122), who in the West became known for his poetry, was also one of the prominent mathematicians of his time, writing a book titled *Algebra*. The study of mathematics during this time was pursued by other important Arab and Persian scholars of preceding centuries: Al-Khowarizmi (9th century) and Al-Biruni (973–1048) who developed a great deal of the theory of algebra. When Omar Khayyam died in 1123, Arabic science was in a state of decline. At Maragha (in present-day Iran), however, there lived in the next century a mathematician of extraordinary talents: Nasir Eddin Al-Tusi (1201–1274), otherwise known as Nasiraddin. Nasiraddin was the astronomer to Hulagu Khan, who was the grandson of the legendary conqueror Genghis Khan and the brother of Kublai Khan. Nasiraddin compiled an Arabic version of Euclid's works and a treatise on the Euclidean postulates. Like the classical mathematicians who preceded him as well as two earlier Arab mathematicians, he too was troubled by Euclid's fifth postulate.

Nasiraddin was the first scholar to recognize the importance of the equivalent postulate to Euclid's fifth: that the sum of the angles of a triangle must be 180 degrees ("two right angles"). Nasiraddin, as did his predecessors, tried to prove that the troubling Euclid's fifth was a mere consequence of the preceding four postulates. And as his predecessors, Nasiraddin failed.

Euclid's classic book was studied widely in the Arab world, leading to discussions about the parallel postulate among other intellectual studies of the book, but Europe did not know it. In the early 1100s, an English traveler, Adelhard of Bath (ca. 1075–1160), journeyed from Asia Minor to Egypt and North Africa and learned Arabic on his way. He then disguised himself as a Moslem student and crossed the straits of Gibraltar into Moorish Spain. Adelhard reached Cordova around the year 1120 and obtained an Arabic copy of the *Elements*. He secretly translated Euclid's book into Latin, and smuggled it across the Pyrenees into Christian Europe. It was thus that Euclid finally arrived in the West, and the book was copied and distributed among scholars and intellectuals who now learned the fundamentals of the geometry the Greeks had known a millennium and a half earlier. When printing became available, one of the first mathematical books to be typeset was the *Elements*. When Euclid's book was published in Venice in 1482, it was a translation into Latin of the Arabic text smuggled by Adelhard. It was only in 1505, also in Venice, that Zamberti published a version of the *Elements* that was a translation of the Greek text, edited by Theon of Alexandria in the fourth century.

Five hundred years had passed since Nasiraddin's work on the fifth postulate, but during these centuries Western mathematics had made little progress. The Middle Ages were not a good time for mathematics or science and culture in general. A world embroiled in constant strife and battered by plagues is not a place for the pursuit of knowledge and art. Then, in 1733 a small book written in Latin was published in Milan. Its title was *Euclides ab omni naevo vindicatus* ("Euclid Cleared of Every Flaw"). The author was a Jesuit priest named Girolamo Saccheri (1667–1733). The book was published in the year its author died, but that was not the only misfortune to society: this groundbreaking book, which could have changed the way people understood geometry for all history, remained hidden for over a hundred years. It was discovered by chance in 1889—after the three people who would change geometry and its interpretation had already published their own mutually independent discoveries. The three were Gauss, Bolyai, and Lobachevsky.

While teaching grammar and studying philosophy at the Jesuit colleges in Italy, Girolamo Saccheri read Euclid's *Elements*. Saccheri was greatly taken by Euclid's use of the method of logical proof called *reductio ad absurdum*. This idea, widely followed in mathematics today, is to assume the opposite of what one tries to prove, then follow logical steps one after the other, until, one hopes, a contradiction is obtained. The contradiction is taken as a proof of the falseness of the original assumption—and hence a proof that its opposite is true, as one had set out to demonstrate.[1] Saccheri was aware of Nasiraddin's work half a millennium earlier and his efforts to derive a proof of Euclid's fifth postulate from Euclid's other four postulates. Saccheri now had a brilliant idea, linking the *reductio ad absurdum* method with the ancient quest to prove the fifth postulate. He decided to attempt a proof by his favorite method. To do so, he had to assume that Euclid's fifth postulate was not a result of the other

1. A simple algebraic example of proof by contradiction is the proof that the square root of two is an irrational number, that is, the square root of two cannot be written as the ratio of two integers. To start, assume the opposite, that is, suppose that there *are* two integers, a and b, whose ratio is equal to the square root of two. Then $a^2 = 2b^2$. Assume, without loss of generality, that the two integers are in lowest terms (they have no common factor, which could be canceled out). If a is odd, we have an immediate contradiction, because $2b^2$ is even. If a is even, it is equal to $2c$, for some number c. So we have $a^2 = (2c)^2 = 4c^2$, which by the assumption must equal $2b^2$, thus b is even and hence a and b have the common factor 2, which is again a contradiction of our assumption.

four postulates, but was actually *false*. By this time, Saccheri was very well-acquainted with Euclid's fifth and with the historical attempts to prove it, since he himself had by then demonstrated that Nasiraddin's proof was false, and that a 1663 attempt, by John Wallis (1616–1703) at Oxford was equally false.

Saccheri assumed that the fifth was false, and hoped to get a contradiction. But he did not obtain one. Saccheri obtained a bizarre outcome: there could be more than one line through a given point parallel to the specified line. Then Saccheri ended up with three possible conclusions, stated in the equivalent form of the fifth postulate, about the sum of the angles of a triangle. The three are that, consistent with Euclid's first four postulates, one can obtain a system where the three angles of a triangle add up to two right angles (Euclid); one where the three angles total less than two right angles (i.e., less than 180 degrees); and one where the sum of the three angles is more than two right angles (i.e., more than 180 degrees). Today, we recognize that the two last systems are separate, *non-Euclidean* geometries, each internally consistent and mathematically valid. They represent views of other worlds. Saccheri obtained a number of important results within these new systems. But he was unaware that he had discovered them and that his failure to obtain a *reductio ad absurdum* was simply because these systems were not false—they were, indeed, mathematically correct! Unfortunately, by the time these facts were recognized by mathematicians, Saccheri would be long dead.

Euclid's fifth postulate, which puzzled and frustrated generations of mathematicians from the day he put it down on paper, encapsulates within it the view that the world is perfectly flat. In such a world, straight lines exist and they can be stretched to infinity, always remaining perfectly straight without bending in the slightest no matter how far they are stretched.[2] Imagine a very flat surface. Here, through a given point, there is one line parallel to a specific line drawn below the point. Parallel lines stretch to infinity and never, ever meet. They remain parallel for eternity. On the flat surface, every triangle has angles that add to 180 degrees. Now think of your flat surface as a flat piece of rubber, and under it a large ball rises, pushing upward. The rubber sheet curves around the ball as it rises, until the surface becomes something like a big balloon. What happens to straight lines that stretch parallel to each other? They bend and on the round surface of the ball they now look like they will meet. On a sphere, there are no non-intersecting lines. And here, the three angles of a triangle add to *more than* 180 degrees. Think of a triangle on a globe with one point at the north pole and the two other points on the equator. Now consider two longitudes. The angle a longitude makes with the equator is a right angle, 90 degrees. So in this triangle on the globe, two of the angles are already equal to 180 degrees. The angle these two longitudes form where they meet at the north pole makes this triangle have the sum of angles over 180 degrees.

Non-Euclidean geometry developed in another direction as anticipated by Saccheri. Our original flat surface was here deformed spherically by a ball pushing up under it. Had Euclid stood on Cape Perpetua and seen that the world is round, and let the concept sink into his consciousness (for he may have known

2. The infinitude of straight lines is implied in Euclid's second postulate. In the late nineteenth century, the great German mathematician G. F. B. Riemann (1826–1866) argued that Euclid's lines could also be interpreted as unbounded and yet not infinite. A great circle on a sphere can be interpreted as a line that is unbounded but finite.

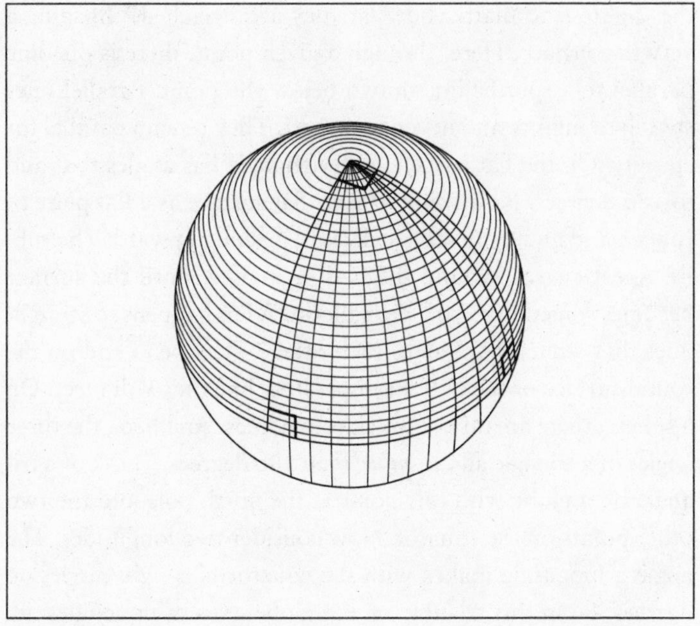

the world is round but not been aware of the importance of this fact), the development of geometry might have been very different. But the flat surface could just as well have been deformed *hyperbolically,* rather than spherically, by stretching it down in the middle and along the contours of a *saddle*. In this world, there are infinitely many possible "lines" parallel to any given line and passing through any specified point not on the line. Here, triangles are thin: the sum of their angles is *less than* 180 degrees.

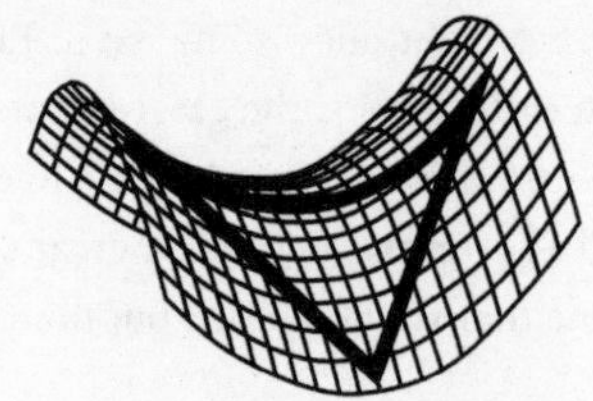

Smaller and Smaller

Smaller and Smaller by M. C. Escher, ©1999 Cordon Art—Baarn—Holland.

It was this strange universe that Saccheri unknowingly entered just before he died. But the important element in both cases, the spherical and the hyperbolic, is that the flat surface has been deformed. Imagine a flat marble table on which straight metal rods have been placed, their edges touching to form triangles. Someone lights a fire under the table. The heat from the fire deforms the rods on the table, and the triangles change: the rods are bent from the heat—and the angles no longer add up to 180

Heaven and Hell

Illustration by M. C. Escher, ©1999 Cordon Art—Baarn—Holland. All rights reserved.

degrees. It was exactly this example that Albert Einstein would give two centuries later to describe how non-Euclidean geometry materializes in the real world.

At the beginning of the nineteenth century, Karl Friedrich Gauss (1777–1855), the famous German genius who contributed prodigiously to science, was the dominant figure in mathematics. Gauss spent decades meditating and pondering the problem of Euclid's fifth postulate. But Gauss published little about the riddle that consumed so much of his time and

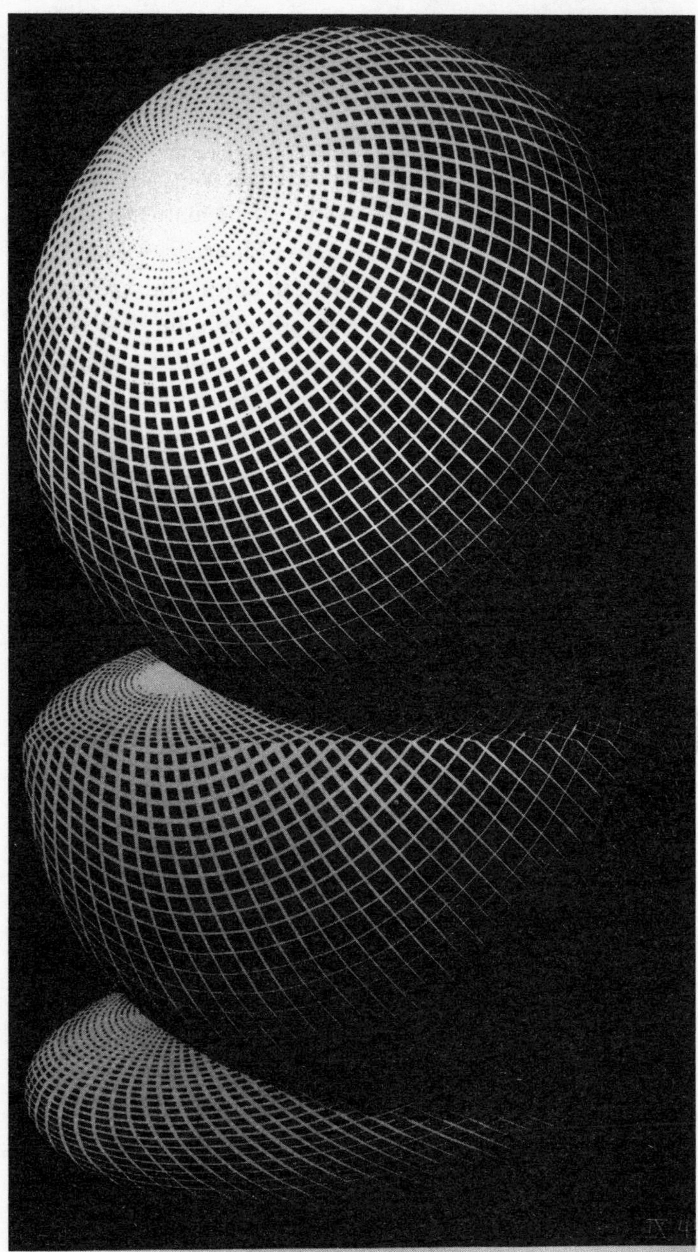

Three Spheres I

Three Spheres I by M. C. Escher, ©1999 Cordon Art—Baarn—Holland.

energy—he wrote about so many important problems in mathematics, but his ideas on geometry we know mostly from his letters. Gauss clearly understood that refuting the fifth postulate leads to non-Euclidean geometries.

While studying at the renowned University of Göttingen, Gauss had befriended a fellow student of mathematics who came to the university from his native Hungary, Wolfgang (or Farkas) Bolyai (1775–1856). Gauss and Bolyai both spent much time trying to prove Euclid's fifth postulate. In 1804, Bolyai thought he had a proof and wrote it as a short manuscript, which he sent to his old college friend. Gauss, however, quickly found an error in the proof. Undaunted, Bolyai continued his efforts and a few years later sent Gauss another attempted proof. This one was wrong as well. While he was a professor, a dramatist, a poet, a musician, an inventor, Wolfgang Bolyai continued his research in mathematics throughout his life despite his failed attempts to prove the impossible postulate. On December 15, 1802, Wolfgang's son was born, Johann (Janos) Bolyai (1802–1860). Wolfgang wrote an enthusiastic letter to Gauss telling him of the birth of his son, "a healthy and very beautiful child with a good disposition, black hair and eyebrows, and burning deep blue eyes, which at times sparkle like two jewels."

Johann grew up and was taught mathematics by his father. He caught his father's preoccupation with Euclid's fifth postulate and the desire to prove that it followed from Euclid's other postulates and propositions. In 1817, the younger Bolyai enrolled at the Royal College for Engineers in Vienna, where he devoted much of his time pursuing his father's passionate goal of proving the fifth postulate. By that time, his father was writing him letters trying desperately to dissuade his son from wasting time on an impossible problem that had consumed so much of his own energy.

The son was not swayed. He was adamant in pursuing his goal, possibly to redeem his father's failed efforts of many decades. In 1820, Johann Bolyai came to a startling conclusion. Far from being provable apart from the rest of Euclid's geometry, the fifth postulate was a gateway to an enchanted garden: a new *Absolute Science of Spa*ce, as he called it, of which Euclid's geometry was just a special case.

Bolyai started with the Playfair version of Euclid's fifth: that only one line can be drawn through a point not on a given line and parallel to it. Bolyai then assumed that the postulate was not true. This would mean, he concluded, that either there was no line that was parallel to the given line, or that there was more than one such line. But by Euclid's other assumptions, a straight line was *infinite*. This implication was shown to contradict the former assumption, leaving the second possibility as the viable alternative to Euclid's fifth postulate. If two lines were parallel to a given line through a point not on the line, then there were infinitely many such lines. This is shown in the figure below.

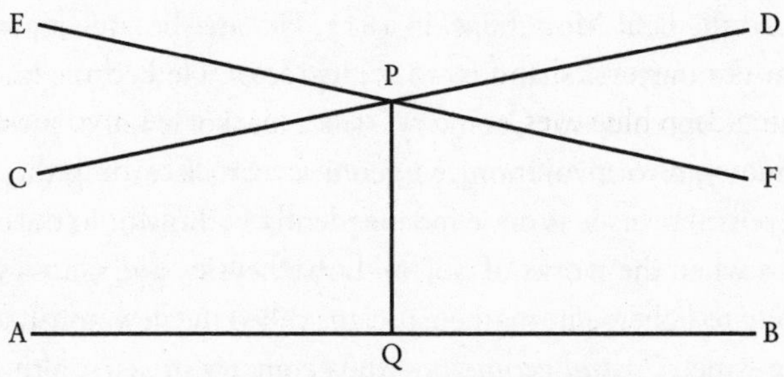

Lines CD and EF are parallel to AB through P.

The results that followed this implication bewildered the young Bolyai. His new geometry developed with no contradictions, no hurdles, as if God Himself had intended the geometry of space to follow this amazing new non-Euclidean path. He noticed with special enthusiasm that there were many propositions that followed without any reliance on any assumption about parallel lines, and which were therefore common to all the possible geometries: Euclidean and non-Euclidean. These propositions contained the essence of the nature of space. Bolyai, only 21 in 1823, wrote to his father that "Out of nothing, I have created a strange new universe."

The father offered to help, and eventually included his son's pathbreaking work as an appendix to his own book on mathematics, whose shortened title was the *Tentamen*, published in 1832.

Gauss, having seen the book by the two Bolyais, commented that he had, himself, reached similar conclusions over three and a half decades of meditating on the problem of the fifth postulate. But yet another mathematician had reached similar conclusions. Nikolai Ivanovich Lobachevsky (1793–1856) took a degree from the University of Kazan, which lies 400 miles east of Moscow toward the Ural Mountains, in 1813. He later became a professor at the university, and in 1827, its rector. He became known as the "Copernicus of Geometry," since his work, Lobachevskian Geometry, also revolutionized geometry by discarding the parallel postulate as was done independently by Bolyai. In the early 1800s when the works of Bolyai, Lobachevsky, and Gauss were coming to light, some mathematicians called the new non-Euclidean geometry *astral geometry*—the geometry of stars, although it is not clear how they came to such a name.[3]

3. In 1818, Karl Schweikart used this term to describe non-Euclidean geometry to his friend Gerling, a professor of astronomy at the University of Marburg, who was a student of Gauss.

In the Bolyai-Lobachevsky-Gauss geometry, the sum of the angles of a triangle is not 180 degrees. And a circle in this geometry is not the usual circle from (Euclidean) everyday life: here, the ratio of the circumference of a circle to its diameter is not equal to the natural number pi.

<>

Einstein followed the line of reasoning that began with the "happiest thought" of his life. Still at the Swiss patent office, he conducted one of his famous thought-experiments. Einstein imagined a circle spinning in space. The center of the circle did not move, but its circumference was moving quickly in a circular direction. Einstein compared what happens in several *reference frames*, a standard tool he had used in developing the special theory of relativity. He concluded, using his special relativity, that the boundary of the disk contracted as it spun. There was a force acting on the circle at the boundary—the centrifugal force—and its action was analogous to that of a gravitational force. But the same contraction that affected the outer circle left the diameter unchanged. Thus, Einstein concluded, in a way that surprised even him, the ratio of the circle to the diameter was no longer pi. He deduced that in the presence of a gravitational force (or field), the geometry of space is non-Euclidean.

CHAPTER 5

Grossmann's Notebooks

"He, on good terms with the teachers and understanding everything; I, a pariah, discounted and little loved."
—Einstein in a letter to Marcel Grossmann's widow.[1]

Einstein, a willful and impatient student, nonetheless needed a solid grounding in mathematics for his revolutionary theories. Much of this he got by cribbing from the notebooks of a better-behaved student, Marcel Grossmann. Marcel Grossmann (1878–1936) was born in Budapest to a family with a long Swiss lineage. When he was fifteen, Grossmann returned to Switzerland, finished high school, and from 1896 to 1900 studied at the ETH in Zürich. Grossmann studied mathematics at the University of Zürich, specializing in geometry, and earned a doctorate in this field. He later wrote papers and textbooks on non-Euclidean geometry.

In contrast to Einstein, his fellow student at the ETH at the turn of the century, Grossmann was very conscientious, always attending classes and taking meticulous notes—a teacher's dream of a student. Grossmann attended the lectures of Minkowski and of other mathematicians and physicists at the ETH. His notebooks—which are now preserved and displayed

1. R. W. Clark, *Einstein: The Life and Times,* New York: Avon, 1972, p. 62.

at the archives of the ETH—were later of crucial importance for Einstein in developing the mathematics he badly needed in order to produce his general theory of relativity. Einstein's key equation was based on this, and other, more advanced mathematics. But Einstein was indebted to his friend Grossmann for more than mathematics. Grossmann's father helped Einstein obtain his position with the Swiss patent office in Bern when the young graduate could not find a job. In 1905, the year Einstein published his first paper on special relativity and the equation $E=mc^2$, he also submitted his doctoral dissertation to the University of Zürich. The thesis, "On a new determination of molecular dimensions," was dedicated to his friend Marcel Grossmann.

It was Grossmann who, in late 1911, contacted Einstein in Prague to find out whether he would be interested in returning to Switzerland to take up a position at the ETH in Zürich where he had been a student. Einstein, who by then had offers from a number of other universities in Europe, was thrilled to accept the offer from the ETH and return to Swiss soil. While he had to take up Austro-Hungarian citizenship in order to be able to accept his appointment in Prague earlier that year, he had also kept his Swiss citizenship. In early 1912, Einstein returned to his beloved Switzerland.

Having concluded that space is non-Euclidean, Einstein needed help. He came for this help to his old friend, now a recognized expert in exactly the area Einstein needed to understand. Some Einstein biographers and writers of books on relativity claim that Einstein was not good in mathematics. There is nothing further from the truth. The scientist who gave the world the theories of relativity was a superb mathematician. The problem was that in his early days as a student at the ETH, Einstein didn't care much about sitting in lecture halls to listen to mathematics. He understood enough mathematics to devise special relativity, and he was able to pick up whatever else he needed on his own. Einstein's relationship with the mathematician Hermann Minkowski serves as a case in point. Einstein did not take seriously Minkowski's lectures at the ETH. Years later, when special relativity was accepted by the scientific community, Minkowski wrote about the mathematics of Einstein's relativity, whose four-dimensional space is often referred to as Minkowski space.

Unlike Einstein, Grossmann was a serious student of mathematics. His notes hold a special place in the development of Einstein's general theory of relativity. Now back at the ETH, Einstein realized that he needed help—very urgent help. If space is non-Euclidean, then he would have to understand its geometry well before he could do anything further with his ideas about gravitation and relativity. For Einstein knew next to nothing about the actual geometries of space.

Grossmann pulled out his yellowing lecture notes from the turn of the century, and looked for hints as to where Einstein should start in his model of the universe and its forces of gravity. The notes, and Grossmann's subsequent work on geometry, told him that the specific methods his good friend would need were those developed in the late 1800s by two Italian mathematicians: Gregorio Ricci, and his gifted student Tullio Levi-Civita. Ironically, back in Prague, the mathematician Georg Pick had also told Einstein that the work of these two scholars could help him develop the mathematics he needed for pursuing his theory. But apparently Einstein was not impressed at the time. Now with Grossmann as his guide to the world of geometry, he was eager to listen.

Non-Euclidean geometry itself could not give answers to Einstein's questions. Such geometries describe space in terms of lines, angles, parallels, circles, and so on. Einstein needed a lot more. He needed, most of all, a quality of invariance. Good physical laws are invariant: they do not change as the frame of reference or the units of measurement change. It should take two hours to drive a distance of 120 miles at 60 miles an hour, and the answer should not change if we denote the distance in kilometers and the speed in kilometers per hour. Einstein was looking for a mathematical tool to allow him to transcend the curvature of space—its non-Euclidean nature—so that the variables of the theory would be valid in any kind of space curvature. Grossmann was generous with his notes and references, but this wasn't enough to solve Einstein's riddle of gravitation.

After working hard on the problem for several months in 1912, Einstein issued a plea to his old friend: "Grossmann, Du musst mir helfen, sonst werd' ich verruckt!" (Grossmann, you must help me or else I'll go crazy!). Grossmann heeded the plea and began collaborating in earnest with Einstein. The result was a number of papers the two wrote on the problem of gravitation. These papers were another step in the direction of a general theory of relativity, but they fell short of what was needed for a complete understanding of the complicated phenomena they purported to describe.

It was then that Einstein turned his attention to the concept of a tensor. This concept also helps demonstrate the increasingly complex mathematics needed to solve the problems of relativity: first the special theory, and then the more complicated general theory of relativity. Simple systems can be described by equations whose elements are single-number variables. A line, for example, is given by the equation $y=ax+b$. Here x and y are single numbers and a and b are coefficients, which are also single numbers. In a line with slope $a=2$ and intercept $b=3$, one can solve the value of y when $x=5$ as: $y=2(5) + 3=13$. As the problem becomes more complicated, one may require several equations, or an equation whose variables are sets of numbers. Here, x would be a *vector*, an ordered set of numbers, and the same would be true of y and any other variables. In physics, velocity, acceleration, and force are all vectors, since all of them have both a magnitude and a direction, and thus each of them is described as a set of numbers.

But what Einstein needed now was a generalization of a vector to yet another level of complexity. He needed a tensor—a variable that is an extension of the concept of a vector. A vector in three-dimensional space has three components. A (second-order) tensor in three-dimensional space has $3^2=9$ components. A tensor maintains the invariance principle required by Einstein, and it accounts for the variable in a complex situation. General relativity indeed posed very complex problems—Einstein had to take into consideration ten quantities, denoted $g_{\mu\nu}$, which accounted for the curvature of the space of four dimensions (three for space, one for time). The animal that accounted for the curvature $g_{\mu\nu}$ was a tensor called the *metric tensor*, since it was a measure of distance in curved space. But the mathematics needed to yield meaningful results was not yet at hand. Something else was needed—something more general than the Ricci and Levi-Civita results. Einstein had to have a way of manipulating the metric tensor so that the invariance principle would hold under any transformation of his equations—he needed a way of transcending the curvature of space, whatever form that curvature might take. His work with Grossmann allowed him invariance only under linear transformations, a situation which was too restrictive for what he had to achieve. But Einstein only became fully aware of the shortcomings of his work with Grossmann in the summer of 1913.

Einstein enjoyed very much his new life in Zürich. He was at a place he knew and loved, and he was with his family. His wife, Mileva, and their two sons were very attached to Switzerland, and this contributed to his sense of well-being. And Einstein was among friends. It was here that he began discussing the problems of the universe with students and colleagues. Einstein's developing equations of gravitation already had some implications about the universe as a whole, and he was enthusiastically trying to explore these implications about the universe in which we live.

Friends and colleagues often described a carefree Einstein leaving the lecture halls of the university, surrounded by a group of students, and heading to his favorite café, the Terrasse Café at the bottom of the Zürichberg. They would spend hours there discussing the philosophical implications of the theories about the extent, shape, past and future of the vastness of space in which we live.

But in the spring of 1913, Einstein had a visit that would change his life and again cause him to uproot himself and his family and move to another country. It was a visit to Zürich by Max Planck (1858–1947) and Hermann Nernst (1864–1941). Max Planck was the greatest physicist of the time—he was a key figure in the development of the quantum theory. And according to a later admission by Einstein, Planck was the only scientist he truly admired. We know that the admiration and respect were mutual. Planck and the physicist Hermann Nernst had lobbied hard in Berlin for an invitation to Einstein to join the faculty of the University of Berlin.

Planck and Nernst arrived in Zürich and met Einstein at his apartment. By that time, he had other offers, among them one for a professorship in Leiden, the Netherlands. The two worked hard at convincing Einstein to take the position in Berlin, but he did not want to make a quick decision. While he was making up his mind, Planck and Nernst went on a mountain-climbing trip in the Swiss Alps. Einstein promised them that by the time they returned, he would have his answer for them. "There will be a sign, so you will know my answer as soon as you see me," he said. When their train rolled into the Zürich railway station, they saw Einstein standing on the platform. He was holding a red rose in his hand.

Einstein's move from the Zürich he loved to a Berlin where anti-semitism was already rising has been a topic of great speculation. It seems that Einstein had several reasons for making such an unexpected decision. First, Berlin was a far more important scientific center than was Zürich. Giants such as Planck lived there. Second, Einstein's position required no teaching. This was an important consideration, since Einstein often complained that teaching responsibilities took too much time and energy away from his research activity. And a third reason was that Einstein wanted to be near a major observatory, so he could interact with astronomers. More than ever, he strongly desired an astronomical proof of the bending-of-light principle of his evolving theory of general relativity. In Berlin, there was at least one astronomer with whom he had been corresponding regularly—Erwin Finlay Freundlich.

Einstein did not immediately perceive a problem with the equations he had developed with Grossmann. In early 1913, he wrote a letter to his friend Paul Ehrenfest (1880–1933), in which he summed up his achievements: "The gravitation affair has been clarified to my full satisfaction. One can specifically prove that generally covariant equations that completely determine the field from the matter tensor cannot exist." But within two years, Einstein recognized his mistake, and in fact developed generally covariant equations—his field equations of gravitation. This happened in Berlin at the height of World War I. But

Paul Ehrenfest, Albert Einstein, ca. 1921
Photograph Willem J. Luyten, courtesy AIP Emilio Segrè Visual Archives

Einstein left behind him in Zürich a curious little notebook containing his derivations of equations and attempts at arriving at the desired field equation of gravitation. This notebook would be discovered by researchers eighty years later and led to unexpected findings about Einstein's work.

Einstein and Grossmann parted ways when Einstein left Zürich. Grossmann spent the following years dealing with social and political issues. He became deeply involved with charitable aid to students of all nationalities who had become prisoners of war. In 1920, he began to show signs of multiple sclerosis, from which he would eventually die in 1936. In 1931, long after Einstein's general theory of relativity had been accepted by the world, Grossmann wrote a bitter treatise against aspects of the theory, apparently in anger after having heard that Einstein had given a lecture on these topics. Einstein seems to have forgiven the betrayal of their friendship and research association, and in 1955 wrote about Grossmann and their collaboration in a moving and affectionate tone. He wrote that he had later discovered that the mathematical difficulty with which he and Grossmann had struggled for many months in their work had been solved almost a century earlier by the German mathematician Bernhard Riemann.

CHAPTER 6

The Crimean Expedition

"I am glad that our colleagues are busying themselves with my theory—even if it's with the hope of killing it."

—Albert Einstein, in a letter to Erwin Finlay Freundlich, August 7, 1914.

The Crimea, August 1, 1914

As Germany declared war on Russia, a German scientist was caught by the Russians by the Black Sea and was transferred to Odessa. Erwin Finlay Freundlich was suspected of being a German spy. He was traveling with strange looking equipment—a telescope. His equipment was confiscated and he was held until the end of August, when he and his team were exchanged for high-ranking Russian officers held by the Germans. Throughout his imprisonment, Freundlich maintained that he was a scientist and that he had come to observe an eclipse. Back in Berlin, Freundlich called on Albert Einstein. Why did Freundlich risk his life to travel to a nation at war with his own? What did he intend to do there? And what was his relationship with Einstein, a German scientist who had renounced his German citizenship only to reclaim it later and move to Berlin?

<>

Shortly after his meeting Pollak in Berlin, Erwin Freundlich began his collaboration with Einstein, who was still in Prague. The two met in Berlin in April 1912 when Einstein worked out the gravitational lensing problem.[1] A year later, during his honeymoon, Freundlich and his bride met Einstein while on their visit to Zürich. As the newlyweds' train arrived at the Zürich station in early September, 1913, they saw Fritz Haber on the platform waiting to meet them. Haber was then the director of the Kaiser Wilhelm Institute, and with him, in untidy sporty clothes and a straw hat, was Albert Einstein.

Einstein invited Freundlich and his bride to accompany him to Frauenfeld, where he was to give a talk about relativity. From there, they traveled to the shore of Lake Constance, and later back to Zürich. Einstein was earnestly discussing the problems of the theory and ways to verify results with Freundlich throughout

1. Actually, the exact place and time of Einstein's first meeting with Freundlich remain a mystery. Ronald Clark, who bases some of his claims about the Freundlich-Einstein relationship in his biography on a conversation with Mrs. Kathe Freundlich, claims that the two men met for the first time in Zürich in 1913 (*Einstein: The Life and Times*, New York: Avon, 1984, p. 207). However, in a letter to Michele Besso, datelined Prague, March 26, 1912 (Document 377, *The Collected Papers of Albert Einstein*, M. J. Klein, et al., editors, Princeton University Press, 1993, Vol. V), Einstein tells his friend that he is soon going to Berlin to meet Planck, Nernst, Haber, and "an astronomer." This astronomer is very likely Freundlich, since he is the only astronomer with whom Einstein had any dealings during this time. Einstein's notebook containing appointments during his stay in Berlin that year has been recovered and studied by Jürgen Renn and his colleagues at the Max Planck Institute for the History of Science in Berlin. While the notebook contains important astronomical ideas as well as the names and times of meetings with various people, no mention of Freundlich is made there. However, the most convincing piece of evidence is a 1935 letter from Leo W. Pollak to Einstein (Document 11-180 of the Einstein Archives in Jerusalem) in which he says that he introduced the two men in 1911.

the entire time. On November 8, Einstein received a letter from a Professor Campbell of the Lick Observatory in California, to whom he had written requesting that photographs of the stars near the sun be taken by the observatory during an eclipse and sent to Freundlich for analysis. The analysis did not bear fruit.

Einstein's relationship with Freundlich is mostly known from a surviving collection of 25 letters that Einstein wrote to the younger astronomer over a period of 20 years, from 1911 to 1931.[2] The letters tell a fascinating story, whose full details have until now not been told. It is a story about the vagaries of fate. It is a story of how the world's greatest theoretical physicist desperately wished for experimental confirmation of his hypothesis and hoped to obtain it through the work of an eager young astronomer. It is a story about the evil of war and the evil of politics and how both stood in the way of humanity's quest for knowledge. And yet it is also a story about luck, faith, confidence, and the mercurial nature of human relationships.

As soon as he received word from Pollak about the interest in his work by the young astronomer, Einstein wrote back to Freundlich. His letter was very polite, almost ingratiating, in its language and especially its address. In his first letter, and in many to follow, Einstein—who by then was already a well-known physicist, if not the world figure he would become within less than a decade—addresses the neophyte astronomer as "Highly Esteemed Mr. Colleague." He then continues to thank Freundlich profusely for his great interest in such an important problem (the theory of general relativity). He encourages him to make every effort to find observational evidence for the predictions of the theory, saying that astronomers can provide a great

2. The collection is kept at the Pierpont Morgan Library in New York.

service to science by finding such evidence. There is a hint of desperation in Einstein's tone, and, reading his letters, one clearly senses that he would do anything to prove by physical means that his theory was indeed right.

Space is curved around massive objects, and a light ray passing by such an object will be bent. In addition, a ray of light climbing up a gravitational field will lose energy, as evidenced by a shift in frequency toward the red end of the spectrum (a gravitational redshift)—just like a person gets tired clambering up a winding stairway. Einstein concentrated his attentions on the light-bending phenomenon he was sure existed in nature. He asked the young astronomer if there was a way of detecting such an event.

In September, Einstein responded stoically to what seems to have been his first disappointment with Freundlich's efforts on his behalf. He wrote: "If only nature had given us a planet bigger than Jupiter!—but nature did not give us the possibility of help in these discoveries." In looking for confirmation of the theorized phenomenon of bending of light rays, it seems that Freundlich chose to look at light passing by the planet Jupiter on its way to Earth. He did not find such an effect. In the hindsight of close to a century, it is easy to see why Freundlich failed. The bending effect is relatively small, and Jupiter—while much larger than Earth—is only one thousandth as massive as the Sun. The planet's mass is not enough to allow measurement of the bending of light rays around it.

On September 21, Einstein had new ideas. He realized that the smallest massive object around which the bending of light had a chance of being detected would have to be the Sun. He asked his Highly Esteemed Colleague what he thought about the possibility of looking for starlight in daytime. Clearly this would be necessary since starlight emanated from a far-away point in space, and if it could be seen in daylight at an angle that makes it pass close to the Sun, bending might be detected if one knew the expected position of the star. One could then compare the expected position of the rays with their observed position, affected by bending because of their close passage to the Sun, and thus establish the existence of the phenomenon. Einstein wanted to know whether astronomers had a way of seeing stars in daylight and finding a star that appears near the Sun's position in the sky.

In early 1913 Einstein wrote Freundlich, again thanking him for his very interesting letter and for his great devotion to the search for proofs of The Theory. He included titillating details of his continuing work on the extended concept of relativity and sprinkled the letter with questions clearly aimed at arousing his younger colleague's excitement for the project. Einstein's letter gives the flavor of his frantic quest for a finished theory. He writes in strong terms about his feelings about competing theories—those of Abraham, Mie, and Nordström. Gunnar Nordström (1881–1923), a Finnish physicist, did some ingenious work on Einstein's field equations. The development of these equations by Einstein and Grossmann ran into trouble because of dependencies of parameters. Nordström's idea was to try to develop an alternative theory of general relativity where the speed of light, *c*, does not depend on a field introduced in Einstein's equations. In his letter to Freundlich, and in following correspondence, Einstein brings out the frantic nature of his quest. This is a fantastic theory, Einstein writes about Nordström's work, but it has a small probability of being right. If Nordström's theory is correct, Einstein writes, then there will be a redshift due to gravitation, but no bending of light. Einstein is desperate, therefore, to find a way of detecting whether light rays get bent in the gravitational field of massive objects: such a test would show whether Einstein—or Nordström—is right. His language leaves no doubt about Einstein's competitiveness. He is convinced that he alone is correct with his (not yet complete) theory.

It is in this letter, written at an uncertain date in early 1913, that Einstein first mentions an eclipse of the Sun. Earlier, in 1912, Einstein seemed to think that starlight could be observed in daylight as it passed by the Sun. Sometime between late 1912 and early 1913, he and Freundlich apparently reached the conclusion that there was no such possibility. At some point it must have occurred to one of them that the phenomenon of a total eclipse of the Sun would offer a perfect venue for the experiment. During a total eclipse it is indeed daytime and the Sun is up, but stars should be observable as well because of the darkness afforded by the Moon's shadow. So while nature did not give us a big enough Jupiter, it did give us this wonderful phenomenon, which occurs roughly every couple of years *somewhere* on Earth and allows us to view stars as well as the exact position of the Sun in the middle of the day.

As soon as he realized this point, Einstein felt great excitement. In a letter he tells Freundlich that he had read in an American journal that several optical systems should be used together in an eclipse in order to see stars around the sun. Here he says that this seems reasonable to his "layman's brain." But in subsequent correspondence Einstein is anything but a layman when it comes to astronomy. Evidently the great theoretician had come to the conclusion that his theory alone would not be worth much without physical verifications. He seems to have taught himself a great deal about astronomy in a relatively short period of time. In many letters he occupies himself with very technical questions about the actual process of designing a system to view an eclipse and carefully preparing the photographic plates necessary for making the photographs of stars in the vicinity of the Sun.

On August 2, 1913, Einstein repeats his conviction: "Theoretically we came to a result—I am quite convinced that the rays of light get bent. I am exceptionally interested in your plan to observe stars in daytime." He then continues with a long dissertation on small particles suspended in the atmosphere and their possible effects on visibility, the quality of the pictures which might be taken, and other technical astronomical points. He then explains to Freundlich "When operating with an optical system, you must get the whole Sun in the picture, together with the part of the sky that interests us—all on one plate. But it has been recommended that two optical systems be used together. What is not clear is how the two pictures will be used together. I would like very much to hear from you what you think about this and any other methods." It seems that Einstein was so determined that everything work right that he was not going to leave to astronomers the details of their everyday work.

And the theoretician was not working on the light bending problem alone. In the same letter he continues saying he is very curious about Freundlich's research on double stars. Freundlich's idea was to look at systems of double stars orbiting each other. Then, if one could somehow estimate the masses of the two stars in a pair as well as their radial velocities around each other, possibly one could detect the gravitational redshift predicted by Einstein's general relativity as the light from one star passes by the other star. This, unfortunately, proved an experimental dead-end. Neither Freundlich's work nor that of others would lead to any result. The phenomenon would be finally detected in an experiment done at Harvard University in the 1960s. But pursuing this particular goal, Freundlich would make serious calculation errors that would eventually irritate

Einstein. The letter concludes with the statement that if such an experiment should somehow lead to a detection of a difference in the *speed* of light (rather than its frequency—as manifested by a redshift), then "the entire relativity theory inclusive of gravitation theory would be false." In closing, he says how happy he would be to finally see Freundlich when he comes to Zürich with his bride on their honeymoon.

By the next letter, sent by Einstein from Zürich on October 22, 1913, they had already met in Switzerland and talked extensively about the problem of detecting the putative bending of light rays from distant stars as they pass by the Sun. After the perfunctory "Herr Kollege," Einstein writes: "I thank you very much from the heart for your extensive news and the deep interest you have shown in our problem." Apparently Freundlich had tried to obtain pictures taken by other astronomers of eclipses past and attempted to discern any images of stars near the shadow of the Sun. In all of these attempts he had failed. It is not hard to understand why this happened. While the Sun itself is completely hidden during a total solar eclipse, the same is not true for the Sun's corona. Bright tongues of fire extend from the hidden Sun to large distances around the circular dark shadow of the moon hiding the body of the Sun. Stars in the area of the corona are very difficult to discern, and detecting their shifts would require an experiment designed specifically for this purpose. Unfortunately no one had performed this experiment before, since no one had perceived a need to detect shifts in stars' positions near the Sun.

From the rest of the letter it becomes clear that the idea of the eclipse was Einstein's and not Freundlich's. In fact, Freundlich's careless nature—as would be evidenced later when he would make elementary errors in computing mass estimates of double stars—comes through in this early context. Einstein spends considerable time in his letter countering Freundlich's apparent contention that detection of starlight shifts near the Sun could be done in daytime without a total solar eclipse. Einstein explains patiently that he had asked local astronomers whether such an endeavor would be possible and was answered with a resounding "no."[3]

By December 7, 1913, Einstein and Freundlich had agreed on the venue for the experiment to verify the bending of light rays around the Sun: an expedition *should* be mounted to go to the Crimea to observe the total solar eclipse predicted for August 1914. They were now deep into the details of the expedition, and the remaining question was how to finance it. Having heard from Freundlich that all was arranged, a complete plan of how to travel to Russia and from there to the Crimea, how to use the telescopic equipment he had devised, how to take pictures of the Sun and the surrounding sky during the eclipse, and how to compare the pictures with those of the same area of the sky taken at night with the stars near the Sun in their usual places, Einstein immediately contacted Planck. He asked him for help in obtaining financial support to try to prove the part of general relativity he felt he had already developed.

It seems that the Prussian academy, however, was not excited enough about the project to fund it. In his letter of December 7 to Freundlich, Einstein says that Planck was interested in the problem, but if the academy would not allocate funds, he, Einstein, would spend his own meager savings on the venture. Einstein, apparently in frustration and anger at not being able to get funding, underlined in his letter: "*I will not write to Struve.*"

3. This feat is still not possible today. Even during an eclipse, detecting the bending of light requires a complicated procedure.

Hermann Struve was the director of the Royal Observatory at Potsdam. Einstein had hoped to secure funding for the project from the Observatory, but apparently he had been rebuffed. Then he added: "If nothing works then I shall pay from my own small savings at least the first 2,000 marks. So please order the necessary plates and let's not lose time because of the question of money."

And then suddenly things began to happen, and nothing could stop the course of events—both of science and of history. On April 6, 1914, Einstein and his family moved from Zürich to Berlin. With Haber's help he found a flat, but within a short time Mileva and Albert separated and she took the children with her and returned to Switzerland. Einstein then moved to a bachelor's apartment and seems to have adjusted to the change, even though it was quite painful as he was very attached to his two sons. He reacquainted himself with relatives in Berlin and found one of them, Elsa Einstein, a cousin, especially pleasant and began to develop a close friendship with her. Within five years, after a divorce from Mileva, the two were married.

On July 2, 1914, Einstein was made a member of the Prussian academy. At 34, he was by far the youngest. All the others were scientists of long standing and more advanced ages. From reports of his conversations with colleagues while still in Zürich before being notified of the honor soon to be bestowed upon him, we know that Einstein did not care much for the distinction. Still, he gave a good speech to the membership, thanking them for the honor and for the freedom membership in the academy would now give him to pursue his research full time. As a member of the academy he would not have to worry about teaching duties or other obligations and would be able to devote his entire time to his research. From correspondence with colleagues we know that Einstein liked living in Berlin and the new status his position afforded him. He was now also in a position to pursue funding for the experimental project with renewed vigor.

Where Einstein had failed while still in Zürich, Freundlich had in the meantime met with some partial success: Director Struve had (grudgingly) agreed to allow Freundlich to take on the eclipse project, but without allocating any observatory funding. Now in Berlin and a member of the academy, Einstein went to work hard on the problem of money. Finally, the academy granted the project 2,000 marks—the money Einstein stood ready to commit of his own savings—and earmarked the funds for the conversion of scientific instruments for the purpose of observing the eclipse, and also for the purchase of the necessary photographic plates. There was still the need for 3,000 marks for travel and freight costs to the Crimea. In one of the many twists of fate in the story of the eclipse, the money came from what in hindsight looks like an unlikely source.

Gustav Krupp (1870–1950) was a German tycoon whose arms manufacturing firm could by that time have been held indirectly responsible for many massacres, including that of the Armenians by the Turks, who used Krupp arms. In 1918, Krupp designed special long-range guns for the purpose of shelling the civilian population of Paris from a distance of seventy-four miles, which would kill 256 Parisians.[4] It was Krupp money that allowed Hitler to campaign against the Versailles Treaty, and in 1933 gave Hitler the necessary votes to gain a majority in the Reichstag and gain absolute control in Germany. The firm supplied the Nazis in World War II, and was an indispensable tool for the Nazi horror. In 1914, Gustav Krupp contributed 3,000

4. Martin Gilbert, *A History of the Twentieth Century*, Vol. I, New York: Morrow, 1997, p. 490.

marks to support the expedition to look for proof for Einstein's theory of general relativity.

As the time of the expedition drew near, Einstein became more and more excited, agitated and withdrawn. His biographer Ronald Clark tells about Einstein's frequent visits to the Freundlich family during the tense period before the planned eclipse expedition. Einstein apparently did not leave anything to chance as he kept his expedition leader, Freundlich, constantly in sight. He often brought his work with him when visiting at the Freundlich house, where, before dinner was over, he would push back his plate and start writing equations on top of his hosts' expensive tablecloth. Freundlich's widow would tell Clark many years later that she regretted not keeping the tablecloth, as her husband had suggested, since it would probably be worth a lot.[5]

It seems that with time Einstein harbored a combination of two feelings. First, he was very anxious about the results of the coming eclipse expedition. He was now a scientist of some renown: his special theory of relativity had been reasonably well-received by the scientific community—although it still had opponents. The general theory, still in its infancy, was getting attention from other scientists and receiving both fierce competition and a fair amount of skepticism. His colleagues at the academy were all older and mainstream in their views and career paths, and among them Einstein was and would continue to be an outsider. He even had to force himself during this period to abandon his natural inclination for careless dress and to don respectable attire as appropriate to his new status. Einstein was

5. Ronald W. Clark, *Einstein: The Life and Times*, New York: Avon, 1984, p.222.

desperate for positive proof that his outlandish theory about space and time and gravitation was correct. At the same time, Einstein was becoming more and more confident about the validity of his theory. In a letter to his good friend Michele Angelo Besso, Einstein wrote: "I no longer doubt the correctness of the whole system, whether the observation of the eclipse succeeds or not. The sense of the thing is too evident." Ironically, Einstein was wrong. As fate would have it, Freundlich would go off to the Crimea looking for light deflection of an amount only half that actually existing in nature. But as Freundlich was about to leave, Einstein had already made up his mind that if the results of the crucial test were to be negative, the experiment would be faulty—not his theory. This would be one of the strongest examples of how in certain cases, to a theoretician, the equation designed to describe nature takes on a life of its own and is viewed as so elegant and so divine that reality takes second place to the formula.

On July 19, 1914, Erwin Freundlich left Berlin with two colleagues, one of them a technician with the famous German lens manufacturer Carl Zeiss. Within a week's travel they reached the town of Feodosiya in the Crimea and prepared their equipment for the eclipse. Freundlich brought four different cameras as well as telescopic equipment to maximize the chances of taking at least one good picture showing clearly stars in the vicinity of the Sun during the eclipse. The German team met another from Argentina, also there to photograph the eclipse for other purposes. Interestingly, the Argentine team came to try to capture on film Vulcan—a hypothesized small planet near the Sun believed to exist because of small systematic aberrations that had been detected for decades in the orbit of the planet Mercury. Vulcan and its putative orbit near the Sun were believed to be responsible for the perihelion problem of Mercury. In another bizarre twist of fate, it would be Einstein's general theory of relativity, which the German team came to test, that would eventually solve the perihelion problem. The shifting of the orbit of Mercury is not due to another planet. None exists. It is due to the effects of the Sun's gravitational field on the planet which lies so close to it. And within a few short years, it would be Freundlich himself who would compile a long list of historical astronomical observations of the orbit of Mercury that, together with general relativity, would solve the problem. Now the two teams, the one from Argentina and the German one, shared information and techniques, as well as equipment in tense anticipation of the Sun's disappearance for a precious two minutes on August 21.

But all this time, history was taking another course—one far away from the direction of science and knowledge. Three weeks before Freundlich's departure from Berlin to the Crimea, Archduke Franz Ferdinand, the heir to the throne of the Austro-Hungarian empire, was visiting the city of Sarajevo, the capital of the province of Bosnia, which had been annexed by the Empire. The Serbian foreign minister took the unusual step of warning the archduke about the visit, advising him that there had been Serbian agitation in the capital and that perhaps now was not a good time for a visit. Franz Ferdinand was undeterred. On June 28, as the archduke's motorcade was carrying him to a ceremony in town hall, a bomb was hurtled at the archduke's car. It exploded, but the archduke and his wife, the Duchess of Hohenberg, were unhurt. But the plot against the Austro-Hungarian prince was thicker—other conspirators waited for him farther down his road. A second bomb was thrown at the archduke, but did not explode. Continuing on its way, the motorcade came to a point where the archduke's car had to pull back and change direction. At that moment, the third conspirator, a nineteen-year-old student named Gavrilo Princip, pulled out a pistol and fired at Franz Ferdinand and his wife, killing them both. The conspirators belonged to a terrorist group called Black Hand, which was actually opposed to the Serbian government. The group's aim was to achieve independence of the south Slavic peoples from the Hapsburg empire. The murder sent shock waves around the world. The storm clouds that would bring the First World War began to thicken over Europe. It seems, however, that despite the anger about the attack voiced both by the Hapsburgs and by their ally, the German kaiser in Berlin, the organizers of the Freundlich expedition to the Crimea were completely oblivious to the political implications of the situation, and never considered that war with Russia might be an imminent possibility. While these mighty forces of war were about to be unleashed, Freundlich and his crew were calmly preparing to watch an eclipse, set to happen on August 21, in Russian-controlled territory.

The kaiser was participating in a regatta in the harbor of Kiel when a note about the assassination of Archduke Franz Ferdinand was thrown onto his yacht folded into a gold cigarette case. The kaiser was enraged and decided to return summarily to Berlin. His ambassador in Vienna suggested that a mild punishment be meted out to the Serbs, but Wilhelm II was implacable. He was determined that the Serbs must be "disposed of—and soon." Public opinion in Germany backed him, and on July 4, the German ambassador to Britain advised Lord Haldane that he was very worried about the developing situation and the distinct possibility of war. Britain advised moderation and hoped for peace. It had nothing to gain from war, and everything to lose. But this was not the case for Germany and Austria-Hungary. The kaiser was especially belligerent because of his feelings about Russia. He was convinced that Russia must be stopped or it would dominate Europe and threaten German hegemony on the continent. The kaiser offered strong support for Emperor Franz Josef in trying to avenge the death of his son on the Serbians—this despite a report of the results of the investigation of the murder of the archduke, which found no involvement by the Serbian government.

On July 23, 1914, the Austro-Hungarian empire issued an official ultimatum to Serbia. The document was unique in the history of nations, for by it Austro-Hungary attempted to dictate to Serbia what to do on an internal as well as external level. There were fifteen demands in the ultimatum ranging from requiring the Serbian government to prohibit all anti-Austrian propaganda within its borders to insisting on the inclusion of Austrian officials in the committee set to investigate the murder. The alternative to full compliance by Serbia was war. The Serbs agreed to all but one of the conditions, but instead of agreeing to negotiate, Franz Josef mobilized his forces and prepared to attack. The Serbs counted on strong Russian support for its Serbian allies, and Germany stood ready to aid the Austro-Hungarians. The Russian czar, hoping to avert a war, made a proposal on July 27 that would have the Austro-Hungarians and the Serbs talking to each other to negotiate an agreement, but the Austrian government rejected the proposal out of hand. With the nations of Europe taking sides with the Russians and Serbs or with the Austro-Hungarians and Germans, if an armed conflict were to erupt, it was clear that it would not be contained and would become a world war.

Early on August 1, Czar Nicholas appealed for the second time to the German kaiser that their long friendship should prevail and prevent bloodshed between their two nations, but Wilhelm wouldn't budge. However, he hoped to confine the conflict to a war in the east only, without an attack on France and the Low

Countries. But his generals already had plans in place for a western front as well. Later that day, German troops crossed the border into Luxembourg and took the village of Trois Vierges. Trying to limit the war, the kaiser ordered his troops to cross back into Germany, but within hours changed his mind and sent the German army back into Luxembourg and on their way to Belgium. In the evening of the same day, August 1, King George V sent frenzied telegrams from London to Berlin and St. Petersburg in a last-ditch effort to stop the impending First World War. But these attempts were all in vain. Late that evening, the German ambassador to Russia went to see the Russian foreign minister at his palace in St. Petersburg and presented him with the German declaration of war on Russia.

As war broke out, the German team led by Freundlich found itself deep inside enemy territory. Since the Germans carried sensitive optical equipment, the Russians could well suspect them of being spies. In the first days of August, 1914, Freundlich's team was arrested. The team members were kept as prisoners of war. On August 4, Einstein, sick with anxiety, wrote to his friend Paul Ehrenfest: "My dear astronomer Freundlich will become a prisoner of war in Russia instead of being able to observe the eclipse. I am worried about him." The prisoners of war were taken from the Crimea to the city of Odessa, where they were kept for almost a month. But fortuitously the Germans had just captured a group of high-ranking Russian officers. The Russians were eager to exchange prisoners of war and the Prussian academy intervened with the German government and secured the release of Freundlich and his colleagues for the Russian officers. By September 2, Freundlich was back in Berlin. But Einstein's hopes for verification of his theory through observation of the eclipse were dashed.

While Freundlich spent the rest of the war years in Berlin,

even working part of the time for Einstein, the relationship between the two soured. Einstein's point of view on it is expressed in a letter he wrote Freundlich on September 10, 1921, saying: "I don't think it would be helpful for us to see each other. I am glad that we are understanding each other more (compare it with 1914). For the whole change in our relationship we can be thankful to the English." Einstein apparently blamed Freundlich, who risked his life and freedom for Einstein's theory, for the misfortunes of war, and it seems that his rancor would last for five years, until an Englishman would succeed where Freundlich was not allowed to tread.[6]

But as fate would have it, by that time, 1919, Einstein's theory of general relativity would be complete and he would have corrected his error about the true magnitude of the deflection of light rays by the Sun's gravity. This correction happened on November 18, 1915, when Einstein announced the expected bending of a light ray just grazing the edge of the sun as 1.75 arcseconds, twice the amount he had predicted in 1914. A philosophical question comes up: What would have happened had history allowed Freundlich to carry out his experiment—detecting a shift of 1.75 arcseconds (plus or minus an experimental error) instead of 0.87 arcsecond as Einstein had predicted (actually, Einstein had made an additional, arithmetic error, obtain-

6. This letter clearly proves that Einstein's relationship with Freundlich started to deteriorate when Freundlich's eclipse expedition failed. Other researchers of the Einstein-Freundlich relationship apparently missed this point. In a recent book, *The Einstein Tower*, Stanford University Press, 1997, pp. 137–8, Klaus Hentschel claims that the relationship began to sour in 1921 when Freundlich attempted to obtain money for one of Einstein's manuscripts, which enraged the latter.

ing the value 0.83 arcsecond). Would general relativity then be judged correct or false?

It should be pointed out that the smaller quantity, 0.87 arcsecond (when computed without the math error) corresponds to the shifting of light when a light ray is considered a particle and when one therefore uses the old Newtonian theory. It was the true incorporation of *relativity* that would lead to a doubling of the value. So quite possibly, had Freundlich's experiment worked, general relativity may not have been considered proven by the scientific community. Perhaps Einstein should have just waited patiently to complete his theory before seeking confirmation, and not blamed his loyal astronomer.

Einstein's ungrateful attitude toward Freundlich manifested itself in many ways over the following years, and one can only feel compassion for the astronomer who had risked so much for Einstein and believed in a theory that had aroused so much skepticism in the scientific community of the time.

Over time, Einstein's dismissive, and at times insensitive, attitude toward the astronomer becomes apparent in his letters. Gone are the "Highly Esteemed Mr. Colleague" addresses in letters Einstein sent him until the failed attempt of 1914. They are replaced by a simple "Dear Freundlich." And in an undated letter of 1917, Einstein writes: "Yesterday Planck spoke with Struve about you. Struve cursed you. You don't do what he expects you to do. Planck thinks that the best solution for you is to get a job teaching theoretical astronomy, and he thinks that you'd have a very good chance of getting one. I think he is right as one shouldn't put all hopes in getting an observatory job. Best regards, A. Einstein."

Einstein continued his correspondence with Freundlich over the years. It seems that he was often asked to help Freundlich find a job or publish a paper. From the tone of letters it is clear that Einstein now felt an important member of the German academic elite, often dropping the name of his renowned friend Planck. Freundlich seems to have been unable to hold positions, and in one letter of 1919 Einstein wrote: "I think a lecturer at the university would be a good position, but it is not easy to get. Don't let this give you gray hair, but enjoy your vacation. All will come to an end. Your nerves are frayed and without a layer of bacon to protect your head. Best regards to you and your wife from our mutual friends, A. Einstein." In another letter, Einstein says he will recommend that the academy accept Freundlich's work *if* Freundlich would answer six technical questions Einstein put to him. On March 1, 1919, by a remarkable coincidence, Einstein wrote Freundlich that he had just read a clear and enjoyable exposition of the work of the English astronomer Arthur Eddington. Unbeknownst to Einstein, Eddington was at that moment about to embark on a trip to an island off the coast of equatorial Africa to watch an eclipse of the Sun and try to detect the bending of star light to prove Einstein's theory of general relativity.

CHAPTER 7

Riemann's Metric

"A geometer like Riemann might almost have foreseen the more important features of the actual world."

—Arthur S. Eddington

Georg Friedrich Bernhard Riemann (1826–1866) was the second of six children born to a Lutheran pastor in the small village of Breselenz in the vicinity of Hanover, Germany. Riemann grew up in modest circumstances and suffered from poor health throughout his short life. It has been said that, had Riemann enjoyed better health and lived even a little longer, the development of several branches of mathematics would have been much accelerated.

Riemann began to show signs of mathematical genius at the age of six, when he not only was able to solve any arithmetical problem posed to him, but to propose new problems to his baffled teachers. When he was ten, Riemann was given lessons in mathematics by a professional teacher who found that Riemann's solutions to problems were better than his own. At the age of fourteen, Riemann invented a perpetual calendar, which he gave as a present to his parents.

Bernhard Riemann was a very shy boy, and he tried to overcome this shyness by overly preparing himself for every public speaking occasion. As an adolescent, he became a perfectionist who would not let go of any piece of work to be seen by others until it was at its best. This propensity to avoid surprises would play an important role in his academic life.

In 1846, the nineteen-year-old Riemann enrolled at the famous University of Göttingen to study theology. His decision was motivated by a desire to please his father, who wanted his son to follow in his footsteps to the clergy. But soon the young Riemann was attracted to the mathematical teachings of the outstanding mathematicians then teaching at the university, among them the great Gauss. With his father's grudging permission, Riemann changed course to mathematics. After a year at Göttingen, Riemann transferred to the University of Berlin, where he received an excellent mathematical education, his mind formed further by the renowned mathematicians Jacobi, Steiner, Dirichlet, Eisenstein, and others. He spent two years at the University of Berlin. Then there were political upheavals in 1848 and Riemann was recruited to serve with the student corps and at one time had to spend sixteen straight hours guarding the King at his palace from violent demonstrators.

In 1849, Riemann returned to the University of Göttingen to work on his doctorate. His doctoral research supervisor was Karl Friedrich Gauss. Riemann made important contributions to geometry, and went on to do work in number theory. Riemann is famous for the *zeta function* he devised, which aids in the study of prime numbers through complex analysis. The problem of finding for which values of a complex variable the zeta function is zero has become one of the most celebrated problems in mathematics. In 1850, after considering problems in many areas of mathematics as well as physics, Riemann came to a deep philosophical conviction that a complete mathematical theory must be established, one which will take the elementary laws governing points and transform them to the great generality of the *plenum* (by which he meant continuously-filled space). This was the idea that would eventually allow him to make a great general breakthrough in mathematics—one that a century later would help revolutionize the future of all of physical science.

In early November, 1851, Riemann presented his doctoral dissertation, with the title *Foundations for the General Theory of Functions of a Complex Variable*, to his advisor Gauss. The work was of such high quality and represented such a substantial contribution to knowledge that Gauss—who had never before done so, nor would ever do so again—praised highly the work of someone other than himself. And this was just a sign of things to come, even though within a few short years both Gauss and Riemann would be dead.

In 1854, Riemann took his first teaching job at the University of Göttingen as an instructor paid for by the students themselves (this was the usual first position of instructors at German universities). It was a custom at German universities at that time to require every new instructor to present a yet-unpublished paper to the department as an initiation rite (such a paper was called a *Habilitationschrift*). Riemann, who had made important contributions in complex analysis and other fields, prepared very carefully and with his usual perfectionism for his presentation. Gauss, the old lion, and every eminent mathematician at the university would all be there to hear him.

Riemann worked incessantly and with his usual fastidiousness. As was the tradition, he had to give the department considering him for the appointment three different topics, ranked in order of his preference, to consider for assigning him as the presentation paper. His first two choices were in his major areas of research, and Riemann had naturally hoped that one of these would be chosen. The third choice was a topic in geometry, for which he was not very well prepared. Typically, the department would assign a candidate his or her first choice, or less frequently, the second, but not the third. So Riemann was working hard on perfecting his preparation for the first two topics.

But Gauss thought differently. As we recall, Gauss had brooded over the problems of Euclid's fifth postulate and the non-Euclidean geometries for many decades, while Bolyai and Lobachevsky developed that field. Gauss, in his meditations on geometry, had developed an idea of curvature. He had defined the *curvature* of the flat Euclidean space as zero, the curvature of a sphere as positive, and the curvature of the hyperbolic "opposite" of a sphere as negative.

Gauss knew the genius of the young Riemann, and thought perhaps *he* could make a breakthrough. He assigned Riemann his third choice.

For the presentation, Riemann developed a whole new theory. Its seeds he had planted earlier. Riemann, while working in the fields of complex numbers and number theory, had spent some of his spare time pursuing his philosophy of space, and had been—on his own—developing the Gaussian idea of curvature, together with those of Bolyai and Lobachevsky. He had a vague notion that a wide, overarching theory lay behind all these disparate concepts of space and its geometry. Was it possible to unify the theory into a powerful new discipline that would transcend detail? This ideal was ever at the back of his mind while he was pursuing problems in other areas. He didn't know whether a generalization was possible until it was almost time for the presentation. The day arrived, and the probationary lecture by a young instructor of mathematics was to be delivered to the older professors. The presentation was a theory that would change the face of both geometry and the physical sciences forever. What was Riemann's groundbreaking idea?

Riemann was one of the finest pure mathematicians of his century. But he was a lot more than a pure mathematician. Deep within his mind burned the desire to understand the nature of the physical world around him. Anticipating relativity and modern cosmology, Riemann understood that in order to grasp the meaning of the physical world, one had to develop a deep understanding of space. And space meant to him *geometry*. Thus, Riemann was interested in describing physical laws as they pertain to the geometry of the space in which we live. He was always a generalist, eschewing details and drudgery in favor of abstraction and generality. Riemann knew that there were three kinds of geometry: Euclidean, hyperbolic, and elliptic or spherical. But he also knew that the geometry of a surface could change in midstream: something didn't have to be only spherical or only Euclidean, for example. A surface could have a geometry that changes from point to point. Riemann wanted something a lot more powerful. He wanted a method that would describe surfaces regardless of how their geometry changed. And this is where Riemann had the revelation that would eventually allow Albert Einstein to complete his theory of general relativity.

Riemann decided that the property of a surface that he needed to understand and capture was the notion of a *distance* (also called a *metric*). In the flat Euclidean space, the shortest distance between two points is the hypotenuse *ac* of the right triangle *abc* if the distance along the *x* direction is *bc* and the distance along the *y* direction is *ab*, as shown below.

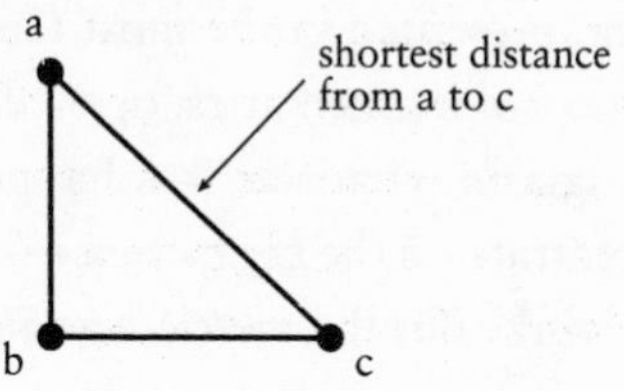

Riemann's genius was to generalize this distance to cases where the space is no longer flat. For example, if the space is curved so that the right angle is no longer right but has magnitude ϕ, then the theorem can be generalized from the Pythagorean $c^2 = a^2 + b^2$ to: $c^2 = a^2 + b^2 - 2ab \cos\phi$. In general, whatever may be the curvature of the space, and even if this curvature changes from point to point along the surface, a function that measures the *instantaneous* distance between two points on the surface was defined by Riemann. The square of the distance function is:

$$ds^2 = g_{\mu\nu} dx_\mu dx_\nu$$

where μ and ν vary over the integers 1,2.

Sixty years later, Albert Einstein would use this very formula, with indices μ and ν varying over the integers 1,2,3, and 4, and accounting for the four dimensions of spacetime (3 for space, 1 for time), to finally derive the equations of general relativity. The term $g_{\mu\nu}$ would be the crucial element in Einstein's tensor equation, signifying the *metric tensor*, which would allow Einstein to account for the curvature that the gravitational field imposes upon the space of the universe. With indices μ and ν varying over the integers 1, 2, 3, and 4, leaving out symmetric terms ($dx_1 dx_2$ is the same as $dx_2 dx_1$), there are 10 quantities that enter the squared distance function definition in four dimensions (why?).

Riemann's idea, presented in the most famous probationary lecture in the history of mathematics, opened a new field. It was now possible to ignore whatever was happening locally on a surface and concentrate on the big picture—since the metric element did all the work. But the metric was also useful in apply-

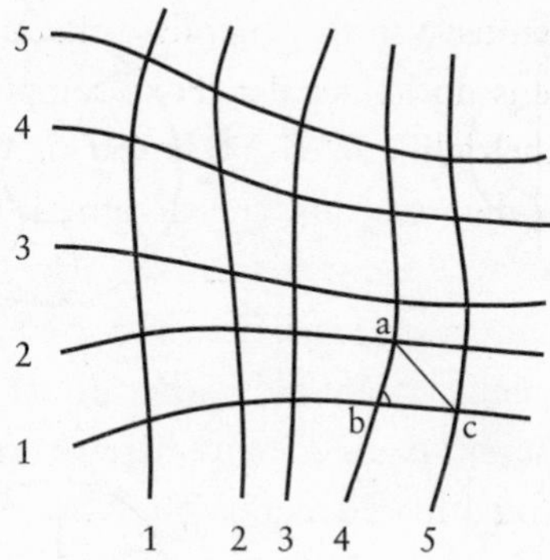

ing methods to the local part of any surface. This led to an entirely new theory: differential geometry, a field which saw great development in the twentieth century. The general approach had implications in the field of topology. Riemann himself studied topological methods in his attack on problems in the theory of functions of a complex variable. Topology is the study of spaces and continuous functions. It deals with questions such as whether a surface is connected or made up of several disconnected components; whether sequences of points converge to a point in the space itself or outside it; and whether it is possible to cover an infinite space with a finite collection of subsets. These are more general questions than those addressed by geometry, but there is a strong connection between the two fields. Topology is also familiar as the field where equivalences are drawn through the use of continuous functions (or deformations)—a doughnut is equivalent in this sense to a cup with a single handle, a sphere is equivalent to any closed three-dimensional surface, and a doughnut with two holes is like a cup with two handles. These equivalences are shown below.

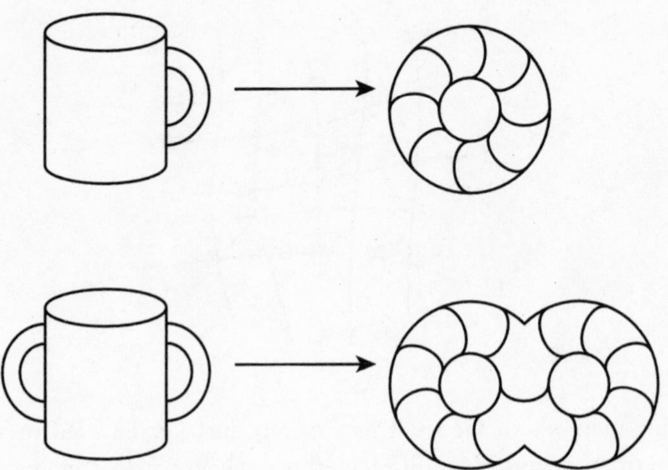

Ultimately, topology says things about the overall geometry of surfaces (also called manifolds). And in this area, grand generalities are possible. By studying topology, mathematicians can arrive at more general and abstract truths than would be possible from geometry. Two well-known examples are the Möbius strip—a two-dimensional surface twisted through the third dimension—and the Klein bottle—a three-dimensional surface twisted through the fourth dimension. The Möbius strip, named after A. F. Möbius (1790–1868), has only one side. This surface is used in moving belts to reduce wear (both "sides" so to speak are used continuously). The Klein bottle is a bottle with no inside. It is named after Felix Klein (1849–1925). Klein was a student of Plucker, who was inspired by Riemann's idea as presented in his *Habilitationschrift* of 1854. Klein had great enthusiasm for geometry and worked to extend geometrical ideas in topology. In this task, he used the powerful algebraic idea of a group. Using group theory, Klein did for topology what Riemann had done for geometry: he achieved unification and abstraction.

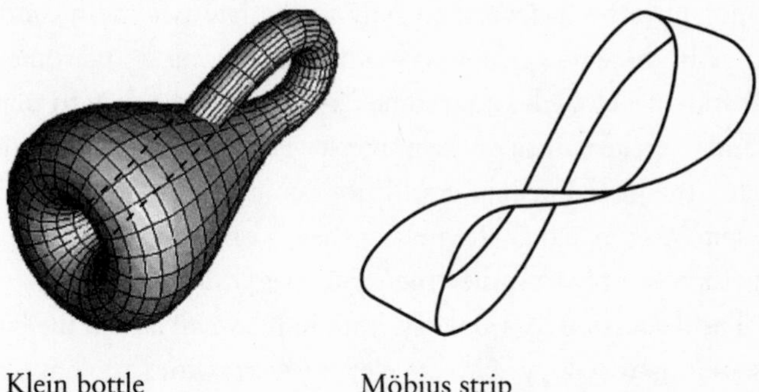

Klein bottle Möbius strip

Riemann's work had, both directly and indirectly, contributed the key ingredients necessary for the understanding of the physical world. His probationary lecture, lauded by Gauss and his colleagues at Göttingen as a great masterpiece, provided the direct and specific tool that would allow Einstein to write down his field equation of the general theory of relativity. Riemann's work in topology, and the inspiration his work provided for Klein and his followers a century later, inspired the British mathematician Sir Roger Penrose to propose an amazing theorem. His theorem, based on Einstein's general relativity, but using the powerful general methods of topology, would establish how our universe must have begun.

Riemann's work in geometry, leading to the modern field of differential geometry, would culminate in results presented at a lecture delivered at Princeton University for the Einstein Centennial Symposium in 1979 (published in 1980) by the world's leading geometer, S.S. Chern. Titled "Relativity and Post-Riemannian Differential Geometry," the lecture contended that the future of general relativity lay in the direction of even more mathematical generality. Chern showed that Riemann's metric could be generalized to more advanced and more complex notions, which have been developed only in the late twentieth century. Possibly, these powerful new mathematical tools—and some not yet fully developed—may some day show us the way to understand the true nature of the universe. They may even allow us to reach the goal Einstein could not achieve in his own lifetime despite his tenacious attempts: a theory of the unification of all the forces of physics, the "theory of everything."

The discussion of geometry, both in the small and in the large, as well as topology with its elegant generalities of shape and space, brings us to an important question: what is the geometry of the entire universe in which we live? Do we live in a giant four-dimensional sphere, or a torus, or perhaps a gigantic Klein bottle? This is one of the most important philosophical questions suggested by Einstein's general relativity and the work of cosmologists in the twentieth century.

In a restricted sense, Riemannian geometry provides a model for the non-Euclidean geometry deduced from an assumption by Saccheri about obtuse angles. A model for this non-Euclidean geometry is the geometry of the surface of a three-dimensional sphere. Here, the sum of the angles of a triangle is greater than 180 degrees. The "lines" in this geometry—the curves of shortest distance between two points on the surface of the sphere—are great circles. A triangle from the north pole to the equator, consisting of two longitudes and the equator, clearly has a sum of angles greater than 180 degrees. A circle in this geometry has circumference less than pi times the diameter. Thus a sphere when viewed as a four-dimensional object provides a model for a non-Euclidean universe of this particular kind. The open, four-dimensional Euclidean space is another possible model for the universe at large. But how do we view a four-dimensional space that is non-Euclidean in the sense of the geometry of Bolyai and Lobachevsky? Here, as noted earlier, the sum of the angles of a triangle is *less than* 180 degrees, and the circumference of a circle is *greater* than pi times the diameter. While the flat Euclidean space has curvature zero, and the sphere or elliptic surfaces have curvature that is a positive number, the curvature of the Bolyai-Lobachevsky geometry is negative. How do we view such a space?

In 1868, the Italian mathematician Eugenio Beltarmi (1835–1900) provided the model for this *hyperbolic* geometry. Beltarmi was also inspired by Riemann's great work, and sought to visualize the space where the Bolyai-Lobachevsky properties are maintained. This surface found by Beltarmi, which has a constant *negative* curvature everywhere, was named by him the *pseudosphere*. It is, in a sense, an inverted sphere, with opposite, negative curvature. The pseudosphere is obtained (in three dimensions) by rotating a tractrix. This is shown below.

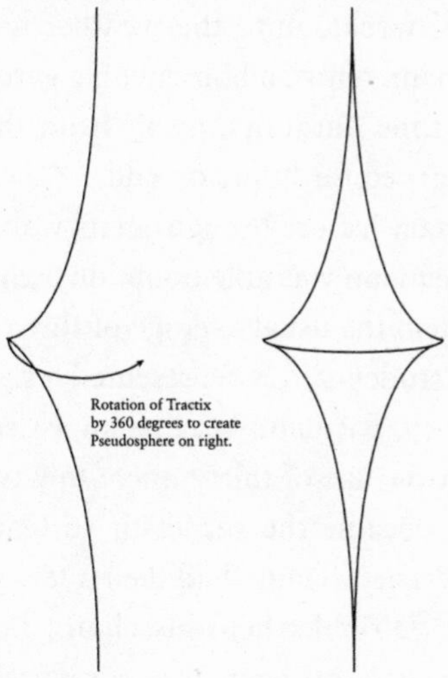
Rotation of Tractix
by 360 degrees to create
Pseudosphere on right.

The geometry of our universe, in four dimensions, is a generalization of one of the three shapes in the figure above. But which one?

<>

Riemann's incredible genius and foresight did not come without a price. Because he was so good, the great Gauss pushed him, and for what he produced under pressure, the world of mathematics—and all of physical science—should be grateful. But the pressure, applied to a person who had a natural inclination to push himself to the limit, along with Riemann's poor health, produced a physical breakdown. Even the exceptionally favorable reception to his *Habilitationschrift*, titled *On the Hypotheses at the Foundations of Geometry*, could not improve his health. Riemann wrote to his father that the extremely difficult investigations he had to carry out both for the probationary lecture and for his ongoing research in mathematical physics and the theory of functions made him ill. He was unable to do any work for several weeks, until the weather had improved. To recuperate, Riemann rented a house with a garden, and made an effort to spend time outdoors, away from the stifling rooms where he had worked for hours on end.

The probationary lecture brought in its wake great academic success. First, Riemann was able to obtain eight students for his lectures rather than the usual average of three or four. Since he was paid by the students, this represented a significant increase in his pay. In 1857, Riemann became an assistant professor at the university, at the age of thirty-one. Only two years later, in 1859, Riemann became the successor to Gauss's prestigious chair at the University. (Gauss had died a few years earlier, and in the meantime Dirichlet held his chair.) Being selected for Gauss's position at Göttingen was a reflection of the great esteem in which Riemann was held by his colleagues, and in fact by the entire world of mathematics.

But his health did not improve, and in 1862 Riemann fell ill again. He had severe problems with his lungs and the German Government granted him funds to travel to the mild climate of Italy for convalescence. Over the next few years, Riemann traveled back and forth between various cities in Italy and Göttingen. When he would arrive at Göttingen, he would fall ill again, while when he was in Italy his health would improve. Aware of his situation, the University of Pisa offered him a professorship, but Riemann declined and made repeated attempts to return to his normal academic life at Göttingen. His condition deteriorated steadily. Riemann died of consumption at a villa on Lago Maggiore in northern Italy in July 1866 at the age of thirty-nine.

CHAPTER 8

Berlin: The Field Equation

"There were two kinds of physicists in Berlin: on the one hand was Einstein, and on the other all the rest."

—Rudolf Ladenburg, one of "the rest."[1]

On July 3, 1913, the Prussian Academy in Berlin voted to accept Einstein for membership in a vote of twenty-one to one. On April 6, 1914, Einstein and his family moved to Berlin.

After his separation from Mileva, Einstein lived in a bachelor's apartment at 13 Wittelsbacherstrasse in the lower-middle-class neighborhood of Wilmersdorf in the southwest part of Berlin. The apartment was on a quiet, tree-lined street, in an architecturally undistinguished multi-storied building. Today, this is a mixed area where foreign-born and native Berliners live. There are flower pots with geraniums decorating most of the balconies, and the cars parked on the streets here are often of the small, old-model, cheap Japanese variety rather than the Mercedes and BMWs one finds in many other parts of Berlin. But no sign on the building indicates that the greatest physicist of the twentieth century once lived there and that in the modest apart-

1. Philipp Frank, *Einstein: His Life and Times*, New York: Knopf, 1953, p. 110.

ment upstairs he worked out the amazing theory of general relativity.

Twenty minutes' walk to the northeast of the apartment lies the fashionable area of the Kurfurstendamm, which Berliners abbreviate affectionately as the Ku'damm. Here, on a wide and busy avenue there were, and still are, shops of every kind and trendy cafés. Farther east, beyond the Ku'damm, lies the sprawling Tiergarten ("animal garden"), one of Europe's largest center-of-the-city parks. When he sought peace and solitude, Einstein could come here and walk among the tall oak trees, or sit by a willow on the edge of the still water where ducks swim and songbirds of many varieties can be heard chirping, and think. He could stroll for hours along the wide paths of the Tiergarten hardly seeing a human being. This was once royal hunting grounds where animals were kept roaming free to be hunted at the king's pleasure.

If Einstein chose to walk through the park east, after an hour or so, he would find himself in the center of Berlin, the Mitte, where he worked in a stately gray stone building, the Preussiche Akademie der Wissenschaften at 8, Unter den Linden, the ritziest address in all Berlin. This is an impressive building with an inner court and a water fountain at its center and benches circling around it. On the right side a plaque briefly commemorates his seventeen-year tenure there. Inside the square inner court, ivy grows on the walls. Today the building houses the Berlin State Library and Prussian Library.

The nearest cafe is two blocks west of the academy, on Unter den Linden, closer to imposing Brandenburg Gate. It is named Café Einstein. But when I asked him about the origin of the name, the waiter assured me: "It has nothing to do with the physicist." In heavily accented German he said that Einstein simply means ein-Stein—one stone. "You see," he said, "the owner said there was only one stone here once, and so he built it into a café." On the other side of the academy building is one belonging to Humboldt University. There, a large decorative plaque commemorates that in that building Max Planck worked out the quantum theory concept named after him, its symbol, *h*, worked into the display.

With fame came increased affluence, and after his marriage to Elsa, the Einsteins moved to another apartment, east of Wilmersdorf in a more central location in a solidly middle-class neighborhood, and settled at number 5 Haberlandstrasse. Today, a newer building replaces the tall house where the apartment was, and there is no sign that Einstein ever lived here. Across the street, a plaque commemorates that Rudolf Breitscheid, a Prussian Minister lived 1932–3. By Haberlandstrasse 5, a street sign commemorates that Jews had their water and electricity cut off and had properties confiscated in 1942. Otherwise, Berlin, where everything is being rebuilt these days, seems more eager to commemorate its victory over communism than anything else in its history.

Einstein slowly recovered from the blow of not having his theory verified by Freundlich's failed expedition to the Crimea. Not having to worry about photographic plates and astronomical details freed him to return to what he did best—theoretical physics. As history would judge, this was the perfect decision. General relativity in 1914 was far from a completed theory. So while a world war was raging in Europe, Einstein peacefully pursued his research. Politically, he could not have been in a worse place. Berlin at this time was rife with hatreds, passions of war, and budding intolerance and antisemitism. As he recalled in later years, Einstein first encountered the specter of antisemitism not at the Catholic school where he was the only Jew, not at the gymnasium, not even in Prague, but only in Berlin. The course of history would make this only the beginning of a long descent of the German capital into the heart of darkness. But still, Einstein lived a surprisingly placid and extremely productive life in the wartime city, and was able to perfect his amazing theory.

Intellectually, Einstein could not have come to a better place, despite politics and war. During his years in Berlin, the physics department had many of the world's greatest brains among its ranks. These included Planck and Nernst, the twin titans of German science; Max von Laue, the discoverer of X-ray diffraction; James Frank and Gustav Herz, who discovered that the impact of high-velocity electrons produces light of certain colors; and Lise Meitner, a Viennese physicist who had made important contributions to the understanding of radioactivity, and who Einstein would say had surpassed Madame Curie in her work. In later years, this exceptional group of scientists was joined by another Austrian, the quantum theory developer Erwin Schrödinger.

Thus, despite Einstein's deep dislike of all things Prussian, his perception of intolerance and antisemitism, and the unpleasant circumstances of war, Einstein's professional life thrived. The members of the physics department would meet in weekly colloquia to discuss interesting new research topics, and Einstein was present at most of these meetings. He would often ask thought-provoking questions and take an active part—all without calling special attention to himself. This remained true of his stay in Berlin even after he became an international celebrity. Despite his sociability, a characteristic easy laugh and good nature, and his participation in events and social gatherings, many described him as aloof. "These cool blond people make me feel uneasy; they have no psychological comprehension of others," he once confided to a friend.[2] But it is easy to understand his behavior. Einstein now was in the final stretch of his long race to develop the general theory of relativity. He had to devote his time to this task, and to keep social obligations to a minimum.

The race to finish the general theory of relativity is a story of mathematical trial-and-error in solving the great puzzle of matter and gravitation, which Einstein performed at an amazing speed during one incredible month: November 1915. In this work, Einstein had to correct the errors he and Grossmann had made in their work of 1913, and to generalize further work Einstein had done with A. Fokker, a doctoral student at the ETH. From July to October, 1915, Einstein was deeply occupied with the serious limitations he perceived in these earlier joint works.

The equations he had developed in an attempt to reach a general relativistic theory of gravitation did not include the possibility of uniform rotations, which physical considerations would have required. Physical laws must be maintained even if the coordinate system changes, thus a rotating system should follow the same laws as a stationary one. A second problem was that the equations did not explain the total observed amount of precession (a tilt in the rotation axis) in the perihelion (the point nearest the Sun) of the planet Mercury. A third problem was that Einstein's proof of a technical detail of the theory—the uniqueness of the gravitational Lagrangian (the kinetic potential)—was incorrect. A fourth problem, of which Einstein was not aware, was that his predicted deflection of light by the gravitational field of the sun was off by a factor of two. All of these

2. Philipp Frank, *Einstein: His Life and Times*, New York: Knopf, 1957, p.113.

problems would be fixed in a whirlwind of mathematical research using Riemannian geometry during November.

On November 4, Einstein presented a paper on general relativity to the Prussian Academy. This was a newer version, where the equations were covariant with respect to transformations with determinant equal to one. This was a technical detail that was more general than the ones employed in the Einstein-Grossmann and Einstein-Fokker papers of earlier years. He confessed to the academy that he had lost all confidence in the earlier equations he had developed and that his earlier proof was based on a misconception. But Einstein recognized that if he wanted equations that truly described gravitation, he would need a far greater generality. At this point, the advance he had achieved from the papers with Grossmann two years earlier was minuscule. What was now needed was a huge jump in the theoretical understanding of the laws of nature, and a corresponding leap in mathematical complexity. Here, Einstein would use the full power of Riemannian geometry as well as the works of the Italian mathematicians Ricci, Levi-Civita, and Luigi Bianchi (1856–1928). The latter had developed tensor identities that were unknown to Einstein—so he would discover them on his own in the course of his intensive month of work.

Einstein was confounded by seemingly insurmountable technical difficulties. On November 11, working feverishly on the problem, he had inadvertently taken a step back. In imposing an unnecessarily strict restriction on his equations, Einstein ended up with the very same equations he had found so inadequate a week earlier. He was back to square one.

On November 18, Einstein imposed a condition called unimodular invariance and did some mathematical derivations that seemed to lead somewhere new. To his great shock and delight, he discovered that the new theory explained precisely the amount of shift in the perihelion of the planet Mercury. His calculations based on the theory agreed with what the astronomers had been observing about the orbit of the planet. Einstein now had one physical confirmation of at least part of his theory. "For a few days, I was beside myself with joyous excitement," he wrote in a letter to Ehrenfest. He would later use a compilation of astronomical results put together by Freundlich in a publication explaining the precession of the perihelion of Mercury using his theory.

But while general relativity—in this penultimate form it had just taken—did explain an important long-standing astronomical puzzle, it would not be, by itself, enough of a discovery to launch relativity onto the world stage. This distinction would belong to the phenomenon that Einstein had long been most concerned with: the curvature of space, causing light rays to bend around its invisible contours. But here, a great theoretical breakthrough was to happen at just about the exact same time that Einstein would explain the perihelion problem. On November 18, Einstein devoted half a page in his paper to announcing a second discovery brought about by his improved equations describing a gravitational field: the bending of light hypothesized seven years earlier should be of a magnitude of 1.75 arcseconds for a ray just grazing the sun, rather than half that amount. Here Einstein used the full impact of the idea that space itself is curved by the gravitational field, rather than using the Newtonian notions of gravity with the added feature of considering a light quantum—a photon—as a particle with mass.

Einstein continued to work feverishly—he was now very, very close to the final equations describing how a gravitational field warps the fabric of space-time. He knew that here he had fierce competition as well as virulent opposition from a number of scientists. These included Max Abraham (1875–1922), who was opposed to relativity on philosophical grounds, Gustav Mie (1868–1957), who developed a different theory that tried to explain how masses and gravitation interact with electromagnetic phenomena, and his staunch competitor Nordström, whose theory bore some similarity to Einstein's but fell short of the full power of relativity and would be shown by history to be insufficient. On August 7, 1914, Einstein wrote to Freundlich, bitterly complaining about some of his opponents and their theories: "Considering the theory of gravitation of Nordström, where the graded widening of the light rays is not important, one also sees it is built on the a-priori four-dimensional Euclidean space, which to my mind is a kind of superstition. Quite lately Mie started a vicious polemic against my theory in which it is quite clear that the previous points of view are clearly visible. I am happy about the fact that our colleagues are busy with my theory, if only with the hope of killing it."

Unwittingly, now a year later, Einstein was creating for himself yet another competitor—a brilliant mathematician with whom Einstein was on excellent terms. This was the famous David Hilbert (1862–1943). On November 7, Einstein sent Hilbert proofs of his paper in which he had derived gravitational equations after recognizing that his earlier methods were faulty.[3] He then continued to send Hilbert papers with the derivations of his equations as they were evolving, and it is known that Hilbert had been present at a talk Einstein gave about his continuing research on general relativity at Göttingen, where Hilbert was a professor of mathematics. Later, Hilbert congratulated Einstein on his breakthrough discovery

3. Abraham Pais, *'Subtle is the Lord . . . '*, New York: Oxford University Press, 1982, p. 259.

of the cause of the perihelion shift. Having seen and heard Einstein explain his work to him, Hilbert then went on to submit his own paper in Göttingen with equations very similar to Einstein's final product. Eighty years later, a scientific committee would unequivocally clear both men from any suspicion of having taken the other's work and passed it as his own. Hilbert's equations would be judged an interesting footnote to Einstein's field equations of gravitation, since Einstein's derivation was consistent and complete and fully correct. Hilbert's work was a partial mathematical elaboration on Einstein's grand scheme.

Hilbert wrote in 1917 that he was greatly helped in his work at Göttingen by Emmy Nöther (1882–1935), who continued to work with him after the development of the general relativity results. But Einstein, too, was aided by the work of Nöther, whom he would eulogize after her death in 1935: "In the judgement of the most competent living mathematicians, Fraulein Nöther was the most significant creative mathematical genius since the higher education of women began."[4]

Einstein had good reason to describe Emmy Nöther in glowing terms, for it was her theorem from which two important consequences of Einstein's field equation of gravitation can be derived. The first one is a conservation relationship for the energy-momentum tensor, T. This is a desirable physical property of the field equation, but Einstein first argued for its veracity using a wrong mathematical coordinate condition. The second consequence is called the contracted Bianchi identities. These identities constitute important technical conditions that are satisfied by the curvature tensor. The identities ensure that general covariance is maintained: the curvature allows physical

4. Ibid., p. 276.

laws to remain unchanged even when our coordinate system is moving. While Einstein derived the relation, which involves the Ricci tensor and the metric tensor, *g*, on his own, the relationship had actually been derived in 1880 by a German mathematician, Aurel Voss (1845–1931), whose proof did not get attention at the time and the theorem was thus re-discovered by the Italian mathematician Luigi Bianchi (1856–1928). Both results are corollaries of Nöther's overarching theorem. It is interesting that Emmy Nöther—directly in the first case and indirectly in the second—thus helped both competitors in the race for the equations of general relativity: Hilbert and Einstein.

In 1997, the dispute over the priority of the field equation reached a definitive conclusion through the research of L. Corry, J. Renn, and J. Stachel, "Belated Decision in the Hilbert-Einstein Priority Dispute," *Science*, Vol. 278, 14 November 1997, pp. 1270–3. The authors conducted archival research that uncovered the hitherto unnoticed original set of proofs of the manuscript that Hilbert sent to the Journal of the Scientific Society of Göttingen, received there on December 16, 1915. This manuscript obviates any possibility that Einstein could have plagiarized Hilbert's work, and further establishes that Einstein derived the equation correctly, while Hilbert's derivation did not result in a correct equation until after Einstein's work had been published. In the proofs, Hilbert states that his equation is not generally covariant, as required of a valid description of the gravitational problem in the context of relativity.

Einstein's field equation has ten components, which can be seen by considering the effects of the four space-time variables on each other, minus the redundant ones (4x4-6=10). Hilbert, however, derived fourteen components for his equation of gravitation, and the additional four are not covariant, as he himself noted. Hilbert needed the four additional components to guarantee the property of causality within his equation. Doing so, he lost the important covariance—which means that his equation was left with an undesirable dependency on the coordinate system—something that a good physical law cannot have. Einstein, on the other hand, got it right with simply ten components. When Einstein's paper was published, Hilbert claimed that no calculation was necessary for obtaining a fully covariant equation with ten parts. Later, he retracted the statement, conceding that Einstein's work was essential.

There was a mathematical trick, which Einstein the physicist knew while Hilbert—one of the greatest mathematicians of all time—did not. This was the contraction of an essential tensor, the Ricci tensor, and adding the resulting trace into the equation. Hilbert came to understand this only *after* he had seen Einstein's paper, having requested it before it was published. So while Hilbert's own paper was published before Einstein's, Hilbert's original manuscript was flawed and was corrected only after he had seen Einstein's fully correct paper. Understandably, Einstein was furious about what he was sure was plagiarism on the part of a colleague he had trusted. His anger, however, subsided enough and on December 20, 1915, Einstein wrote to Hilbert:

"There has been a certain resentment between us, the cause of which I do not want to analyze any further. I have fought against the feeling of bitterness associated with it, and with complete success. I again think of you with undiminished kindness and I ask you to attempt the same with me." With the Corry, Renn, Stachel article the problem was laid to rest. The theory of general relativity is Einstein's and Einstein's alone.

<>

Historians of science faced a daunting task trying to reconstruct Einstein's stunning discovery of the laws of general relativity.

While editing *The Collected Papers of Albert Einstein*, the continuing multi-volumed project undertaken by John Stachel and his colleagues over the last few years, researchers came upon a notebook containing handwritten notes of Einstein, which he used between summer 1912 and spring 1913, during his time in Zürich. In 1984, Stachel's collaborator John Norton published a paper—based on an analysis of the Zürich Notebook—which corrected misconceptions about Einstein's tortuous path to general relativity. However, some parts of the notebook remained obscure to modern researchers. In 1997, Jürgen Renn and Tilman Sauer undertook a systematic analysis of the Zürich Notebook in an attempt to understand it completely. The notebook, with a title page "Relativität," contains 84 pages of short notes, equations, and calculations with little or no explanations. The analysis, therefore, had to rely on a fundamental understanding of the actual physics and mathematics that underlies the equations. The researchers went through Einstein's jottings one by one, trying to reconstruct his thought processes. The analysis revealed an unexpected result: in 1912, Albert Einstein had written down an approximation of his final field equation of gravitation, formally derived by him three years later. How did this happen?

Einstein's early steps are unsure. He starts by writing the Riemann distance element but he is unused to the notation, using a capital "G" for the metric tensor, later in the notebook changing over to the standard lower-case "g." Then Einstein explores various equations and mathematical manipulations trying to fit together the metric tensor, gravitational elements, and the framework of four-dimensional special relativity. And suddenly it's there: a linear form of the actual field equation of gravitation. But apparently Einstein thought the equation was not correct—there were terms he found disturbing, as they did not satisfy all the requirements. He gave up and spent the following years on a pursuit that led to one blind alley after another. In 1915, probably without being aware he had the equation in his hand three years earlier, he derived it again—this time in its full form, satisfying all his conditions.

In his derivations, Einstein made use of Riemann's metric tensor—the measure of distance in a curved space: $g_{\mu\nu}$. His equations would also include an energy-momentum tensor, T, the Ricci tensor, R, which handles the curvature of spacetime, as well as Newton's gravitational constant, G, and the numbers 8 and π. While these form a set of equations, because tensors contain several elements each, condensing the notation into tensors allows one to write a *single*, and mathematically concise and elegant, equation. Einstein was manipulating these tensor and scalar (numbers, not tensors, such as 8 and π above) in his head and on paper while still intoxicated by his discovery of the explanation for the problem of the shifting perihelion of Mercury. The final stretch would take him all of one intense week. By November 25, 1915, Einstein had in front of him the final equation describing space and time and the resulting curvature due to gravity, with all its implications. He wrote down:

$$R_{\mu\nu} - 1/2\, g_{\mu\nu}R = -\, 8\, \pi\, G\, T_{\mu\nu}$$

On March 20, 1916, Einstein sent to the journal *Annalen der Physik* his paper presenting the systematic derivation and exposition of the complete theory of general relativity. Later in 1916, this paper was expanded and became Einstein's first book.

An equation is a set of symbols and numbers arranged on two sides of the equal sign (=). The equation describes a relationship that is believed to hold among all the quantities represented in the equation. But an equation alone does not provide solutions. The equation must be *solved*. Once Einstein produced his (tensor) equation connecting the various quantities that should determine properties of nature, the task remained to solve the equation. This, in a sense, was Einstein's challenge to the world of science—solve my equation and learn something about the laws of nature. Solving Einstein's field equation means finding the *metric* that satisfies the equation. It means determining the quantity ds^2 of the Riemann metric that applies to any situation as dictated by the equation.

Once one obtains the "line element" ds^2 , one knows the *shape*, the curvature of spacetime in the particular situation to which Einstein's equation is applied. One knows how the "lines" in this curved space look and what constitutes the shortest path between two points. For example, if one applies Einstein's equation to a sphere, one finds that a shortest-distance "line" in this space is a great circle, as every airplane pilot or ship's navigator knows. The equation, of course, is applied to more complicated spaces where gravity plays a role. It accounts for how gravity itself imposes curvature on the space.

The process Einstein set in motion by proposing his equation continues to our own day—and intensifies all the time. Solutions of Einstein's equation have led us to discover fantastic phenomena that the equation predicts. These include gravitational waves, the warping of space, the phenomenon of "frame dragging," where space-time is whirled around a spinning massive body, not to mention the perihelion problem, the gravitational redshift, and the bending of light. But the first solution to Einstein's equation (except for his own solutions leading to the redshift, bending of light, and other effects Einstein determined on his own) was provided even before the theory was complete by a soldier in World War One.

On January 16, 1916, and again on February 24, 1916, Einstein read before the Prussian Academy two papers written by

Karl Schwarzschild (1873–1916), a brilliant German astrophysicist and Director of the Potsdam Observatory, who was the first to solve Einstein's field equation of gravitation. Schwarzschild's solution later led to the understanding of black holes, and ultimately dictated the strong implications of Einstein's field equations on cosmology. Schwarzschild could not read his own papers in front of the academy, because he was at that time languishing in the trenches of the Eastern Front of World War I. There in the battlefields, facing the Russians, Schwarzschild had read Einstein's paper with the equations and solved them. He mailed the solution to Einstein in Berlin. Schwarzschild died on May 11, 1916, from a disease he contracted on the front. On June 29, Einstein read an obituary of Schwarzschild in front of the Prussian Academy.

On May 5, 1916, Einstein was elected to succeed Planck as the president of the German Physical Society. His colleagues in Berlin and elsewhere in the world of science had by now begun to regard him with very high esteem. His derivation of the laws of general relativity were theoretically perfect and the quality of the work did not escape the attention of scientists. But Einstein still did not have what he wanted the most: an observational proof that light bends around the Sun. This final step was needed to transform his work from an elegant theory to an actual description of the laws of the universe. For this he would have to wait another three years. In the meantime, Einstein continued his work with amazing productivity. He wrote a paper describing the phenomenon of gravitational waves. Einstein solved his equation and found out that gravity itself would produce waves that cannot be seen or felt, but which could possibly be detected by extremely sensitive instruments. Many scientists have spent much time, effort, and resources to try to detect gravitational waves. We are close to such a discovery as instrumentation gets better and we begin to use space as our laboratory.

In July, 1916, Einstein returned to work on problems in quantum mechanics. In a few months, he wrote three papers on the subject, in one of them presenting a new derivation of Planck's law. It was during this time that his work in the area of quantum mechanics first brought him the distinct qualms he would have throughout his life about the probabilistic aspects inherent in the quantum theory. His unease led to his oft-quoted statement that "I shall never believe that God plays dice with the world." In December, Einstein was appointed to the Board of Governors of the Royal Physical Technical Institution by decree of the Kaiser. He held this position until he left Germany with Hitler's rise.

CHAPTER 9

Principe Island, 1919

"The description of me and my circumstances in the *Times* shows an amusing flare of imagination on the part of the writer. By an application of the theory of relativity to the taste of the reader, today in Germany I am called a German man of science and in England I am represented as a Swiss Jew. If I come to be regarded as a 'bête noire' the description will be reversed, and I shall become a Swiss Jew for the German and a German for the English."

—Albert Einstein in a letter to the London *Times*, November 28, 1919.

The war made the exchange of information among scientists very difficult. Soon after Einstein published the full theory of general relativity, the Dutch astrophysicist Willem de Sitter (1872–1934) obtained a copy of the paper. He knew that across the English Channel, another bright astrophysicist would be delighted to read Einstein's fantastic theory—and, more than others, would actually understand the intricate details of Einstein's masterpiece. But what was to be done? The war was raging and transporting a document to England would not be easy. De Sitter devised a secret scheme and managed to smuggle Einstein's paper into England and all the way to London where it finally reached its destination: Arthur Eddington (1882–1944).

Arthur Stanley Eddington was born on December 20, 1882, in Kendall, Westmoreland, England, where his father was the headmaster of the local school. When the boy was only two years old, his father died and the mother and her two children moved to the town of Weston. As Eddington was growing up he showed a great fascination with large numbers.[1] At a very young age he learned the 24 x 24 multiplication table by heart. The fascination with large numbers was, in part, what brought him to astronomy. When lecturing, he would often write large numbers on the board with all their digits (rather than using the customary scientific system using exponents). His biographer Chandrasekhar recalls that in a lecture at Oxford in 1926, Eddington wrote on the board the estimated mass of the Sun, in tons, as: 2,000,000,000,000,000,000,000,000,000.

Eddington was educated at Owen's College in Manchester, where he graduated in 1903, and proceeded to Cambridge to study for a doctorate. In 1907 he received the Smith Prize and was elected to a Fellowship at Trinity College. That same year, by the invitation of the Astronomer Royal, Sir William Christie, Eddington joined the staff of the Greenwich Observatory. In 1912 he was elected to the Plumian Chair at Cambridge University. In 1914, Eddington also became the Director of the Cambridge Observatory. He would hold both of these respected positions for the next thirty years. As soon as Eddington received Einstein's paper he was engrossed with it. It was written in language he understood well.

From neutral Switzerland, Einstein used Swiss channels to send over to England copies of his paper, but apparently the copy sent by de Sitter to Eddington was the only one to arrive in Britain until after the end of the war. Incidentally, de Sitter sent to Eddington and to the Royal Astronomical Society three of his

1. S. Chandrasekhar, *Eddington: The Most Distinguished Astrophysicist of His Time*, London: Cambridge University Press, 1975.

own papers on general relativity. One of these papers, on cosmology, would have a decisive impact on Einstein's work on cosmology, and on the course taken in the field of cosmology for decades to come. Eddington did his best to disseminate Einstein's magnificent theory in Britain and the United States.

Eddington prepared a paper titled *Report on the Relativity Theory of Gravitation*, which he published in London in 1918, and which received wide circulation in scientific circles in the West. Eddington was an enthusiastic convert to relativity. As a theoretician of great talent, he immediately recognized in the theory its elegance and its logical foundations. Eddington—the astronomer—saw no reason to look for physical justifications for the beautiful equation of the laws of nature. To him the equation made perfect sense. Ironically, it was this very person, Arthur Eddington, with his great confidence about the veracity of general relativity, who would use physical measurements to prove it to the world. But the idea for the proof came from another source, and was due to a serendipitous succession of events in wartime Britain.

Eddington was a Quaker, and thus—like Einstein—a pacifist. In 1917, England raised the maximum age at which it drafted people into its forces to 35, because it needed more people for the fighting. Eddington, who was 34 years old at that time, was to be inducted, but it was clear he would refuse to go, as he was a conscientious objector. This presented a messy problem for the dons of Trinity College. If Eddington refused the draft, he could be arrested and sent to an internment camp in northern England, where many of his Quaker friends were spending the war years peeling potatoes. This would have caused a great embarrassment to the College, the Observatory, and to British science. Something had to be done, and quickly—before Eddington was called for duty.

British astronomers and physicists who learned about Einstein's paper from Eddington were also aware of the possibility of testing the prediction of the theory by observing the bending of light during an eclipse, as Freundlich had tried to do just at the beginning of the war. In March, 1917, the Astronomer Royal at the time, Sir Frank Dyson, pointed out to scientists that on May 29, 1919, there would be a total eclipse of the Sun. This eclipse was not observable in Europe, but would be observable in a band over the Atlantic Ocean including an area in Brazil as well as the small tropical island of Principe, off the West African coast. Dyson noted that the eclipse would be especially favorable for testing for possible bending of starlight by the rim of the eclipsed Sun because of the constellations involved. During this eclipse, the Sun and Moon would be in the heart of the constellation Taurus, having the rich star cluster of the Hyades right in the center. Such an excellent opportunity shouldn't be missed, he thought.

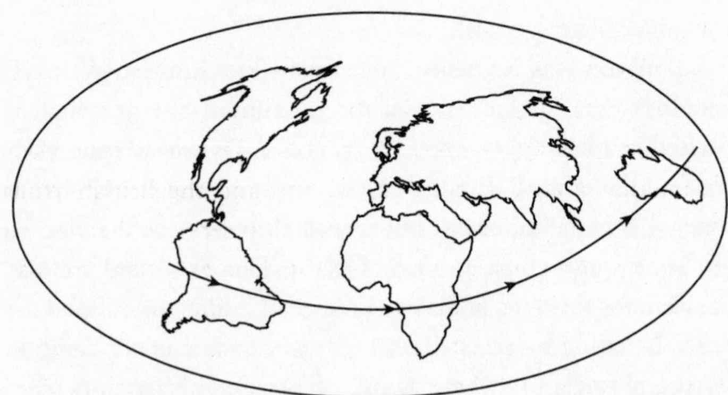

But the world was at war, and sending a British expedition all the way to the equatorial African coast or to Brazil would be hard. Even if the war ended, such an expedition would be difficult to carry out, and the risks and hardships would still be substantial. But there was an adventurous pull to such an idea—mounting an expedition to a faraway and potentially dangerous place in uncertain times for the sake of knowledge. Dyson was intrigued. And he still had that other persistent problem to contend with: Eddington and his conscience. Dyson devised a truly devious scheme—one that would solve both problems at once.

As Astronomer Royal, Dyson had good connections at the British Admiralty. He turned to the admiralty with an unusual request. First, he explained to them the great importance to science of the general theory of relativity (derived by an "enemy scientist"—Einstein had to re-assume his German citizenship upon his move to Berlin) and the incredible opportunity that its verification in the occasion of the eclipse might offer to science. He then argued that Arthur Eddington was the one and only scientist in the West who could accomplish such a verification. It is likely that Dyson had used some patriotic arguments as well: In his book *Opticks*, Sir Isaac Newton asked cryptically: "Do not bodies act upon light at a distance and by their action bend its rays?" Apparently, the problem of whether light is bent by bodies was something that the great English scientist of centuries past wanted to know. In any case, Dyson was successful in making a deal with the Admiralty and the British armed forces: Eddington would not be called to serve in the war, but he must prepare an expedition to the tropics to test general relativity during the eclipse. Furthermore, should the war end in time before the day of the eclipse, May 29, 1919, then—as a national service to Great Britain in lieu of military service—Eddington would have to lead the expedition. The deal was done, and Eddington was not called for military duty.

On November 11, 1918, the armistice agreement stopped the bloodshed of World War I. Then on January 18, 1919, the Paris

Peace Conference was opened at the Quay d'Orsay and the representatives of the nations that took part in the war began to work out their agreements. The war was over, and the expedition to measure the bending of light during the May eclipse could go ahead. In preparation for the scientific enterprise, the Astronomer Royal had obtained funding from the British Government to support the effort, and he and Eddington spent many days going over all the details of the planned expedition. Arthur Eddington would later describe the period of preparation for the trip as the most exciting period in his life. And indeed, how many astronomers or other scientists get to prepare to go to a secluded and mysterious tropical island at a time of worldwide austerity in the aftermath of a great war?

Poring over maps of the expected path of the eclipse, Dyson and Eddington decided that two teams should go—to two different locations—in order to maximize the chances that at least one team would get favorable weather (eclipses can be clouded over, especially in the tropics) and that at least one good photograph showing some stars close to the Sun might be obtained. The path of the eclipse of 29 May, 1919, would take it diagonally from southwest to northeast across the Atlantic Ocean. One of the two locations, therefore, was in Brazil, and the other across the Atlantic by the African coast. The two locations that the astronomers found to hold most promise were Sobral, in a desolate part of the state of Ceara in the Amazon region of northern Brazil, and, across the ocean, the island of Principe. Each location was handled by the staff of an observatory. Brazil would be the destination of a team from the Greenwich Observatory, headed by A. C. D. Crommelin, and Principe would be the responsibility of a team from the Cambridge Observatory. This team would be headed by Arthur Eddington.

The two expeditions left the port of Liverpool together on board H.M.S. Anselm on March 8, 1919, and headed for the island of Madeira in the North Atlantic. There, the two teams parted ways. The team headed for Brazil continued on board the Anselm, arriving at Para on March 23. There they had a choice: they could either continue on to Sobral or wait and arrive there a few weeks later, on the Anselm's return part of the voyage en route back to England. Since they found themselves in an undeveloped area of the Amazon jungle, the team chose to wait on board the ship. By the time of the ship's return to Para, there would be instructions waiting for them with advice on how to proceed, telegraphed to them by a certain Dr. Morize, who had some connections and influence with the Brazilian Government. The telegraphed message and instructions, as well as introductions to a government official, allowed Crommelin and his team to get help in dealing with the local customs officials—especially since they arrived with baggage containing heavy equipment.

At Camocim, the tired astronomers and technicians were met by a local official who helped them with arrangements to have the equipment loaded onto a train which then took them through the Brazilian jungle to Sobral. Natives surrounded the train as it pulled into the station. The local people looked curiously at the British scientists in their colonial jungle outfits as they unloaded strange contraptions from the train. Civil and religious leaders of the region appeared to greet the foreigners. Soon they were joined by Colonel Vicente Saboya, the Deputy of Sobral. He welcomed the visitors, shouted orders to the local porters who were carrying their equipment, and led the entire group to his house.

Soon, local carpenters were hard at work cutting trees, sawing wood, and preparing large wooden V-shaped supports, rest-ing on strong wooden trestles. These were to hold the tube of the telescope at the correct angle to the horizon so that at the appointed time, seven stars could be seen along with the eclipsed Sun. The site for the telescope was right in front of the house given the team by Colonel Saboya. Huts were being built in the mud, as well as a pier, for use by the astronomers. Much time and effort were then spent on adjusting the instruments and checking them. To account for the change in the declination of the Sun during the eclipse, special grooves were made in which the wooden V-supports of the tube could slide, allowing for the change in azimuth. The team used a tube 19 feet long and a lens 16 inches in diameter. The photographic plates to be used to record the eclipse were 10 by 8 inches in size.

To focus the telescope, the astronomers used the bright red star Arcturus. A series of exposures was made, with the focus of the cobalt-glass eyepiece varying slightly from exposure to exposure over the entire range. Then the photographs were carefully examined to find the best focus position to use during the precious moments of the coming eclipse. Once the focus was determined, the breech end was securely tightened to avoid any chance of subsequent movement.

The British team at Sobral then occupied itself compulsively with studying and analyzing the weather. This would be extremely important since the weather could make or break the entire enterprise. The team had apparently chosen its location for the telescope wisely, even though at first sight it might have seemed silly to place a large telescope in front of a house in a settled area of the jungle, since only six miles to the northwest stood the 2,700 foot high Mount Meruoca—still in the path of the eclipse—and it would have seemed logical to place a telescope on top of a high mountain. But as it turned out, the scientists had soon discovered that the mountain was a great point of attraction for clouds in the area, and its summit was frequently enveloped by mist. Down by the town of Sobral, however, the good weather kept the scientists and their telescope dry most of the time.

The temperature was uniform, varying daily from about 75 degrees at 5 AM to about 97 degrees at 3 PM. They also noticed a curious constancy in barometric pressure from day to day. On May 25, however—only four days before the eclipse—there was a heavy rain. The scientists wondered anxiously whether this was a shift in the weather or a passing storm. In the meantime, they welcomed the rain, which moistened the ground and eliminated much of the dust in the air. As they waited, a daily crowd of natives surrounded their compound. They were told that the strangers had come there to see the Sun turn black and the day into night. Would such a terrible thing really happen? They looked on with awe as the scientists milled around the strange tube ominously aimed at the sky.

Arthur Eddington and his team, headed for Principe Island, parted from their colleagues on the H.M.S. Anselm when it reached Madeira and left the ship. They stayed on the Portuguese island for several weeks, waiting for the ship that would take them onward to their destination. The scientists spent their time hanging around their hotel in the city of Funchal, and visiting the lush pineapple plantations and fishing villages on the mountainous green island. So far, if this was how he was to serve his country instead of being bombed in the trenches of World War I, Eddington wasn't doing too badly. On April 9, the cargo vessel *Portugal*, belonging to the Companhia Nacional de Navegacao, came into port and the British team boarded it as it sailed south to the island of Principe, lying one degree north of the equator off the coast of Equatorial Guinea in West Africa. At that time, Principe and the neighboring island of São Tome were both a Portuguese colony.[2]

In the early morning of April 23, 1919, the *Portugal* entered the small port of St. Antonio, on Principe Island. As the ship approached the shore, the passengers saw what looked like a tropical paradise. The beach was lined with fan palms, the sand was white, and the color of the water varied from turquoise to jade. (Many decades later, this perfect beach would be the site of the filming of a famous Bacardi commercial.) Behind the beach, the travelers could see the terrain rise, covered by the canopy of a lush rainforest, farther climbing onto two volcanic mountains, their peaks surrounded with clouds that looked purple in the rising sun. As the ship approached the dock, they could hear the sound of many birds on trees in the thick forest nearby—the island has 26 indigenous species and 126 other recorded kinds of birds. Around them, they could see dongos—oca-wood canoes returning home from the night's fishing, loaded with marlin, sailfish, and barracudas. The fishing grounds were so rich that the islanders used un-baited ropes frayed at the ends to simply tangle the fish and haul them up. This place was paradise.

But Principe Island had a secret. It had only recently ended a centuries-long tenure as a slave colony. The slaves were kept in inhuman conditions, and made to work in the cocoa and banana plantations in the interior of the island, where thousands of them died from hard work, starvation and disease. This was

2. The islands finally gained their independence from Portugal in 1975. The place of the colonial Portuguese government was taken by a one-party system with close ties to Cuba, China, and the Soviet Union. Only recently, with the fall of Communism, did the islands become democratic—one of the smallest independent countries in the world.

one of the last places on earth to ban the slave trade in the nineteenth century. Another beach, which the travelers would later see, was named Stupid Beach—because the Portuguese thought that escaped slaves who settled on the beach would never survive. The island had remnants of its cruel past. And it had poisonous snakes and malaria.

As they disembarked ship and watched their equipment being unloaded by local stevedores, Eddington and his team were met by a delegation of Portuguese colonial officials. All of them pledged their support and offered help to the scientists. It was Vice Admiral Campos Rodrigues, of the National Observatory of Lisbon, who had made the connections for the British team with the local officials who were now ingratiating themselves with the British scientists. In a gesture of good will, the Portuguese government dispensed with the required customs examination of the visitors' voluminous baggage.

The team used some of the first four-wheel-drive vehicles to traverse the ten-by-six mile island in search of a good site to observe the eclipse. They looked at a number of cocoa plantations in the interior of the island, owned by Portuguese colonials and worked by black natives, some of them freed slaves or the descendants of slaves. After a few days of surveying and driving through the thick jungle, Eddington decided on a site: Roca Sundy, in the northwest of the island, overlooking the sea from a height of 500 feet.

The team's heavy baggage was then brought to Sundy from Santo Antonio on April 28. It was transported by vehicles most of the way, but the last kilometer was impassable to the heavier vehicles carrying the equipment. In the thick of the woods, on muddy, mosquito-infested terrain, the heavy equipment was unloaded and carried on the bare backs of many natives. Right in front of the house in which they stayed, the team erected a walled enclosure where they placed their telescope on ground sloped steeply toward the sea in the direction at which the Sun would be during the eclipse. This allowed an unimpeded view of the sky. After a week spent in hectic preparations—local carpenters preparing similar V-shaped supports for the telescope, weights, adjustments as were done about the same time across the ocean at Sobral—the team returned to Santo Antonio. They waited there, spending the week of May 6–13, as Eddington had decided that it was unwise to unpack the telescope's mirror too early in the damp climate. When the team returned to Sundy on May 16, the first check-plates were taken to test the performance of the telescope and photographic equipment. Eddington wanted to leave as little to chance as possible. May 29, eclipse day, would be a defining day in his life.

For the hundredth time, Eddington went over the possible outcomes of the experiment, should nature allow it to proceed without interference of clouds or bad visibility. There were three distinct possibilities as outcomes: 1. Nothing would be found—no change in the apparent positions of the stars during eclipse. This would mean that light does not bend. 2. There would be a shift in the position of the stars—a bending of the light—but in the amount predicted by Newton's theory, applied to light as if it were a particle. 3. There would be the full bending of the light of the stars to the degree predicted by Einstein. The first possibility would mean there is no effect. The second would mean that, with the additional assumption that light can be viewed as a particle, Newton would be the winner, and with him England. The third possibility was that not Newton, but Einstein and his new revolutionary ideas about physics and nature would win. Eddington knew that his compatriots in England hoped that the second possibility would be true. But he rooted for Einstein. He loved the theory of relativity—he understood it, and to him, as to Einstein thousands of miles away and with no idea about the portentous preparations being made in Principe and Sobral, God just *had* to run the universe according to general relativity. Eddington waited for the fateful day. He hoped and prayed for good weather.

The days preceding the eclipse were very cloudy. On the morning of May 29, there was a very heavy thunderstorm from 10 AM to about 11:30 AM—an unusual occurrence for that time of year. Then, an auspicious sign—the Sun appeared for a few moments. But soon the clouds gathered again. The calculated time of the eclipse was 2h. 13m. 5s. to 2h. 18m. 7s. G.M.T. (one hour later, measured in local time). By 1:55 PM (GMT), the crescent Sun could be seen on and off as clouds passed over it. By then, the typical nature of light—and with it the eerie feeling experienced by eclipse observers—was fully perceptible. The light from the mostly-covered Sun took on a translucent quality. It was as if the landscape were seen through a screen. The screen was slowly becoming more opaque. But the clouds were still drifting overhead. Then, just before totality, the clouds parted enough to uncover the Sun. Suddenly, a big shadow raced over from the direction of the water and engulfed the viewers. The total eclipse had begun. Looking up, the observers were stunned by the power of nature. Even veteran eclipse-viewers are moved every time they see a new one. Between the clouds, which parted just long enough for the photographs to be made, the astronomers and their helpers looking up to the sky could see the dark disk of the Sun. Around the disk was the bright halo of the solar corona, burning as a flame, and beyond it darkness as if it was night all the way to the dim red horizons resembling a darkening sunset.

The star field around the Sun was clear, and the photographs (taken at Principe and Sobral) showed a total of thirteen stars.

These—the Hyades—included the relatively-bright (fourth magnitude) stars Kappa Tauri and Upsilon Tauri, along with eleven other dimmer stars.[3] These are shown below.

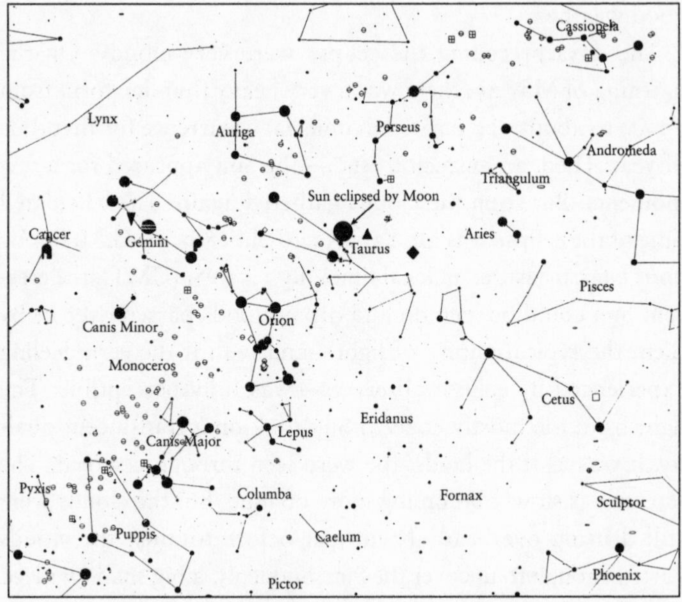

The exposures were made precisely according to the rehearsed plan, using sixteen plates. Cottingham gave Eddington each plate and operated the driving mechanism. Eddington changed the dark slides.

3. Because of an anomaly in their historical development, star magnitudes increase with decreasing number, down to negative numbers. The brightest star, Sirius, is roughly of magnitude -1. Vega, a bright star of the next magnitude down, is magnitude 0. Dimmer stars are of magnitude 1, then 2, and so on. A star of magnitude 4 is barely visible to the naked eye, but is very well observed with even a weak telescope.

Meanwhile, across the Atlantic at Sobral, the team there enjoyed excellent visibility, with no clouds and no agonizing worry about the weather. The astronomers and the local onlookers stared at the powerful spectacle, many of the natives awestruck by an event they had never seen before. Here, photographs of the same stars were taken, but not developed on site. These plates were shipped to England and arrived there long after the team at Principe had developed theirs. The work of the Sobral group therefore served only as a confirmation of the results reached by Arthur Eddington and his team.

When Eddington and his crew on Principe developed the photographs they had a scare—the first ten photographic plates showed no stars at all. In the excitement of the awesome event, heightened by the historical significance it would have if the results were positive, the scientists did not quite notice during the darkness of totality that thin clouds were covering the overlapping Sun and Moon for much of the time. Of the remaining six plates taken, two showed five stars each, which was just enough to get a result. Check-plates for comparison of the stars' positions had been taken at Oxford some months prior to the expedition, showing the same star field including the Hyades and other stars in Taurus. To check for any systematic optical errors, the team also took check plates of another part of the sky, including the bright beacon Arcturus, as was done at Sobral as well.

In his makeshift laboratory by the telescope, the excited Eddington developed and compared the plates of the Hyades star field taken at Oxford and at Principe. The results were stunning: an average displacement of 1.6 arcseconds plus or minus a standard error of 0.3 arcsecond. Within statistical variation, the results matched extremely well the prediction of Einstein's general theory of relativity (a deflection of 1.75 arcseconds). He quickly sent an excited telegram to England: "Through clouds, hopeful. Eddington."

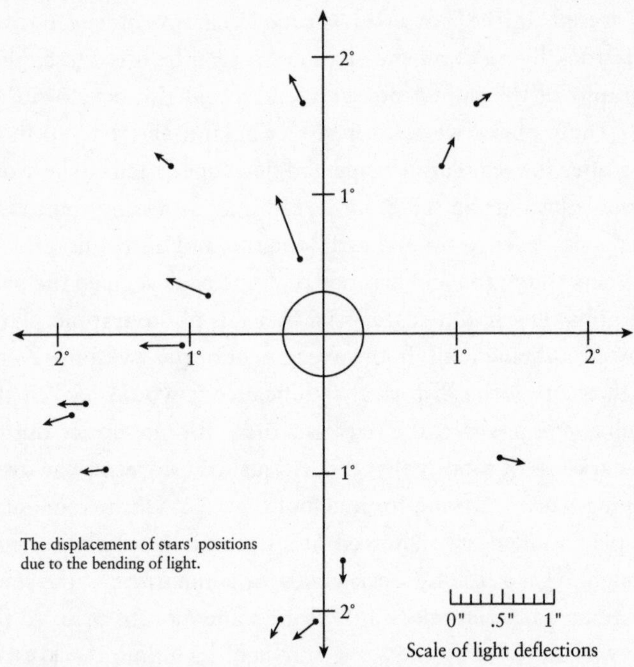

The displacement of stars' positions due to the bending of light.

We know something of Eddington's true state of excitement from the language he used to describe the event in prose and poetry upon his return to England six weeks later (and later recalled in his book):[4]

> Our shadow-box takes up all our attention. There is a marvelous spectacle above, and, as the photographs afterwards

4. Sir Arthur Eddington, *Space, Time and Gravitation: An Outline of the General Relativity Theory*, New York: Harper & Row, 1920, reprinted 1959, p.115.

> revealed, a wonderful prominence-flame is poised a hundred thousand miles above the surface of the sun. We have no time to snatch a glance at it. We are conscious only of the weird half-light of the landscape and the hush of nature, broken by the calls of the observers, and beat of the metronome ticking out the 302 seconds of totality.

Eddington and his team folded up their operation and began packing in preparation for the return to England. They had planned to stay longer to analyze the photographs more completely, but bad news was conveyed to them by their hosts: a strike at the steamship company was imminent. If they didn't want to remain marooned on the island for many months, they had better leave immediately. The island's administrator intervened on their behalf, and at his behest, the Portuguese Government commandeered berths for Eddington and his team aboard a crowded steamer headed for Lisbon. The ship left Principe in a hurry on June 12, just before the strike was to begin. On July 14, the team arrived at the port of Liverpool.

At Sobral, the team stayed on for an additional seven weeks to obtain good reference photographs. On July 18, the team began dismantling their instruments and packing at a much more leisurely pace than had been done at Principe. They left Sobral on July 22, leaving the packed cases with their local host to send on to England. They arrived in England several weeks later. The results of the Sobral team's computations showed an average deflection of 1.98 arcseconds and a standard error of 0.12 arcsecond. These results, too, were within statistical variation in confirmation with Einstein's prediction.[5]

5. Assuming the usual statistical two-standard-deviations rule for significance at 95 percent.

Before they left for Principe, the Astronomer Royal had explained to Eddington's assistant Cottingham, who unlike Eddington was not an expert on relativity, the main idea for the experiment and its great significance. Cottingham got the impression that the bigger the result, the greater would be the excitement, but apparently carried the argument too far—beyond the amount of deflection expected from Einstein's theory. "What would happen," he asked, "if the deflection is even double *that* amount?" "In that case," answered Dyson, the Astronomer Royal, "Eddington will go mad, and you will have to come home alone."

But all ended better than even the optimistic Eddington had expected. Einstein's prediction was verified to within experimental error, and everyone was home. But what about Einstein himself? It was, after all, *his* theory. When would he be given the exciting news?

CHAPTER 10

The Joint Meeting

"Your excellent article in the *Frankfurter Zeitung* gave me much pleasure. But now you, as well as I, will be persecuted by the press and other rabble, although you to a lesser extent. It is so bad for me that I can hardly come up for air, let alone work properly."

—Albert Einstein, in a letter to the physicist
Max Born, December 9, 1919.[1]

In June, 1919, Einstein returned to Berlin from his stint in Zürich. Einstein was vaguely aware that British attempts were made in May to verify the predictions of his general theory of relativity during the eclipse. But Einstein had not been notified that the expeditions had taken place as planned, that results had been obtained, and that these results confirmed his theory. The man who had spent so much time and effort studying areas outside his main interest—practical astronomy, meteorology, and other areas that would pertain to observing stars during an eclipse, and who had courted Freundlich and other astronomers, all in the desperate hope of proving his theory—was simply not informed. While Einstein had made sure his work would be made

1. Max Born, *The Born-Einstein Letters*, New York: Walker, 1971, p. 18.

Conseil de Physique Solvay/Metropole Hotel

available to scientists in Britain, an enemy country, the same British scientists never let him know, in times of peace and free flow of information across borders, that *his* theory had been confirmed. In fact, the British *never* did make the news available to the father of relativity. It was only as late as *September* of 1919, after he inquired in desperation of his friend Lorentz in Holland, that Einstein got the second-hand news that Eddington and his collaborators had proved him right.

Einstein had three good friends in Holland: Lorentz, de Sitter, and Ehrenfest, who was much younger that the other two, about Einstein's age. Back in 1911, at the Solvay Congress in Belgium where many eminent scientists had come to discuss Einstein's relativity, Lorentz hatched a plan to convince Einstein to take a position at the University of Leiden in Holland. This would have brought him in close daily contact with his three friends and disciples of his theory. Einstein declined Lorentz's offer and opted instead to go to Berlin, where there were physicists he considered to be of greater stature, such as Planck.

Undoubtedly, Einstein regretted turning down his friend, who in the following years continued to work on Einstein's theories of relativity, as did de Sitter and Ehrenfest. Throughout the decade, Lorentz, Ehrenfest, and de Sitter kept up their research—trying to follow Einstein's moves as he kept modifying and improving his equations. Einstein himself would describe his struggle as: " . . . my errors in thinking which caused me two years of hard work before at last, in 1915, I recognized them as such and returned penitently to the Riemannian curvature, which enabled me to find the relation to the empirical facts of astronomy."[2]

Letters went back and forth between Ehrenfest, Lorentz and de Sitter on every aspect of Einstein's developing theories. One wonders whether Einstein would have had better—and earlier—success had he decided to go to Leiden instead of Berlin. At least with respect to cosmology, Einstein would have done much better had he kept in better contact with de Sitter than with Planck back in lofty Berlin. At any rate, now in 1919, it was the 66-year-old Lorentz who first brought him the word about Eddington's success. Einstein was intoxicated with delight. On September 27, having just gotten the news, Einstein wrote to his mother: "Dear Mother! Today happy news. H. A. Lorentz

2. Einstein, Albert, *The Origins of the General Theory of Relativity*, Glasgow, U.K.: Jackson, Wylie, 1933.

telegraphed me that the English expeditions have really confirmed the deflection of light by the Sun."

Back in England, the British were moving ahead with the business of general relativity without the slightest concern about Einstein. It was all their affair now: their scientists, their expeditions, which made general relativity a reality. It was now time to announce the discoveries to the world, and discuss the results, debate them as if they were a political issue to be dealt with in a parliamentary system. In November of 1919, a historical joint meeting was held in London of the Royal Astronomical Society and the Royal Society. Here, the pros and cons of the theory of general relativity, as well as the interpretations of the results of the two expeditions of May, were discussed, dissected, embellished. But Einstein himself was not present, and as far as we know he was not even invited. Ironically, this meeting in London launched Einstein to the status of international celebrity as the greatest scientist of the century. It would be these events that transformed him from a physicist to a world personality active and revered not only in science but also in world affairs.

On November 6, 1919, Sir Joseph Thomson, O.M., P.R.S., the chair, called the meeting to order. He then called on the Astronomer Royal to give the assembly a "statement of the purpose and the results of the Eclipse Expedition of May last." Sir Frank Dyson took the stand, and in a long declaration explained how the idea began and how the expeditions were organized, and then gave a summary of the results. He continued: "The effect of the predicted gravitational bending of a ray of light is to throw the star away from the Sun. In measuring the positions of the stars on a photograph to test this displacement, difficulties at once arise about the scale of the photograph. The determination of the scale depends largely upon the outer stars on the plate, while the Einstein effect causes its largest discrepancy on the inner stars nearer the Sun, so that it is quite possible to discriminate between the two causes which affect the star's position."[3]

After the very long address by the Astronomer Royal, which included all the details of the results of the observed shifts in the positions of stars detected both at Sobral and on Principe, it was time for Crommelin, the head of the team that went to Sobral, to give a statement. He started by saying that he didn't have much to add to what the Astronomer Royal had already said, except to express his thanks to the Brazilian authorities for their extensive help to the team, which he did by mentioning by name every Brazilian official who had given help in some way. He thanked steamship personnel, interpreters, meteorologists, laborers, deputies. And then came Eddington. Here the boring details and the senseless pomp suddenly disappeared. Eddington, possibly the only person there with a clear and thorough understanding of the general theory of relativity, treated the assembled members to a tour de force.

After the short perfunctory descriptions of how and when, Eddington went on to discuss the curvature of space that had just been proven by the expeditions. He said that the results clearly pointed to the larger of the two possible deflections of light—the one predicted by Einstein rather than the one that could be assumed to be derived from Newton's laws. He continued: "The simplest interpretation of the bending of the ray is to consider it as an effect of the weight of light. We know that momentum is carried along on the path of a beam of light. Gravity in acting creates momentum in a direction different to that

3. "Joint Eclipse Meeting of the Royal Society and the Royal Astronomical Society," *The Observatory: A Monthly Review of Astronomy*, Vol. XLII, No. 545, November, 1919, p. 389.

of the path of the ray and so causes it to bend." To further explain the difference between the bending of light caused by Newton's law and the full observed bending, of twice the amount, as predicted by Einstein's theory, Eddington actually went on to produce the two geometrical *metrics* of space. Leaving out the time dimension, and concentrating only on space, the distance elements in each of the two theories are given by the metrics as:

Newton's Law: $ds^2 = dr^2 - r^2 d\Theta^2$

Einstein's Law: $ds^2 = -(1-2m/r)dr^2 - r^2 d\Theta^2$

The additional term, $(1-2m/r)$, where m is a mass of a particle and r and Θ are polar coordinates of space, is what differentiates the two measures of distance. The Einsteinian distance is the one that actually measures the warping of space—its non-Euclidean nature—around the massive object, the Sun. In the next meeting of the Royal Astronomical Society, in December, Eddington put this fact about space and its curvature in these words: "There are difficulties in reconciling these results with the laws of Euclidean geometry, but that means that we must choose some law of geometry that will work."[4]

The meeting became a heated competition between two points of view: the Eddington-Dyson view that general relativity had been proven, at least in its aspect of predicting the bending of light to the full extent Einstein explained, and a contrarian point of view held mainly by Sir Oliver Lodge. The latter had originally wagered against the results the two teams now presented.

4. "Meeting of the Royal Astronomical Society, Friday, 1919 December 12," *The Observatory: A Monthly Review of Astronomy*, Vol. XLIII, No. 548, January 1920, p. 35

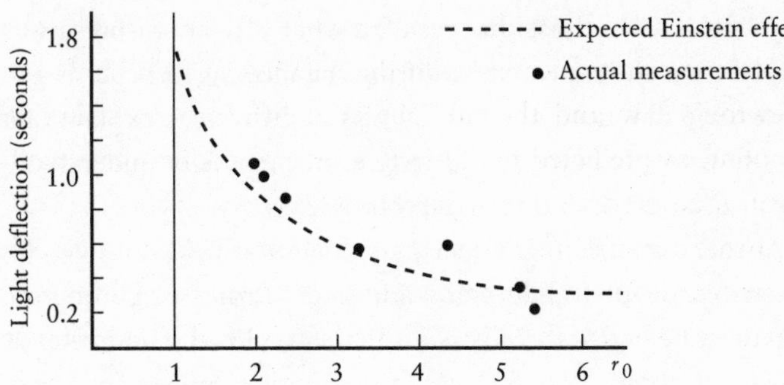

He was a stubborn believer in the old theory, tenaciously holding on to the myth of the ether and other physical theories now overthrown by general relativity. Another skeptic was Ludwig Silberstein, who argued that Einstein's theory was not proven because the gravitational redshift effect had not yet been discovered. Eddington, however, patiently explained that the weight of the present result was overwhelming and that the redshift was another story, for another experiment.

In the end, it was clear that the Eddington-Dyson point of view had prevailed. Space was curved and general relativity was supported. Sir J. J. Thomson, the President of the Royal Society, summed up the majority opinion when he declared at the end of the historic meeting: "This is the most important result obtained in connection with the theory of gravitation since Newton's day, and it is fitting that it should be announced at a meeting of the Society so closely connected with him. If it is sustained that Einstein's reasoning holds good—and it has survived two very severe tests in connection with the perihelion of Mercury and the present eclipse—then it is the result of one of the highest achievements of human thought. The weak point in the theory is the great difficulty in expressing it." The famous late astrophysicist S. Chandrasekhar recalled what J. J. Thomson saw as the "difficulty." Apparently, during the meeting, a belief began to develop that general relativity was so difficult to explain that very few people in the world were even capable of understanding it. Chandrasekhar tells the following story.

At the after-meeting dinner party that was held, Ludwig Silberstein came up to Eddington and said: "Professor Eddington, you must be one of three persons in the world who understands general relativity." On Eddington's hesitation, Silberstein urged, "Don't be modest, Eddington," and Eddington replied: "On the contrary, I am trying to think who the third person is."[5]

At the after-meeting dinner hosted by the Royal Astronomical Society, Eddington read the *Rubaiyat*-like stanzas he had written to commemorate the great success of the eclipse expeditions.

The Clock no question makes of Fasts and Slows,
But steadily and with a constant Rate it goes.
And Lo! The clouds are parting and the Sun
A crescent glimmering on the screen—It shows!—
It shows!

Five minutes, not a moment left to waste,
Five minutes, for the picture to be traced—
The stars are shining, and coronal light
Streams from the Orb of Darkness—Oh make haste!

For in and out, above, about, below
'Tis nothing but a Magic *Shadow*-show

5. S. Chandrasekhar, *Eddington: The Most Distinguished Astrophysicist of His Time*, New York: Cambridge University Press, 1957, p. 30.

Played in a Box whose candle is the Sun
Round which we Phantom Figures come and go.

Oh leave the Wise our measurements to collate.
One thing at least is certain, LIGHT has WEIGHT
One thing is certain and the rest debate—
Light-rays, when near the Sun, DO NOT GO STRAIGHT

The meeting got the full attention of the British media, and on the next day, November 7, the news broke. A headline in the London *Times* proclaimed: "Revolution in Science—New Theory of the Universe—Newton's Ideas Overthrown—Space 'Warped'." The *New York Times* followed some time later, and then newspapers and magazines throughout the world. Within days, Einstein became one of the greatest celebrities—possibly *the greatest*—the world has ever known.

Other eclipse expeditions took place over the years, and all have confirmed the results of the Eddington-Dyson effort of 1919. The next eclipse expedition of 1922 gave very good results in favor of Einstein's theory, and so did every reported eclipse expedition—except for one. At a meeting of the Royal Astronomical Society of January, 1932, Erwin Freundlich, by then an astronomer in Scotland, gave the results of *his* eclipse expedition. He claimed that the deflection of light he had detected was substantially in excess of Einstein's prediction.[6] No wonder that the extensive correspondence between Einstein and Freundlich, extending over a period of twenty years, abruptly comes to an end in 1932. Einstein himself was faced with the embarrassing

6. Ibid., p. 30.

task of defending his theory against his erstwhile friend's faulty conclusion. In a letter to L. Mayr of Gmunden (perhaps a layman who has heard of Freundlich's results in the media), written from his country house in Caputh on April 23, 1932, Einstein says the following.[7] "The results of Mr. Freundlich are based on a faulty calculation of the experimental outcomes (as Mr. Truempler of the Lick Observatory clearly proves in a yet-unpublished paper). With the correct computation, good agreement with the theory results."

Apparently Freundlich's reported results were not taken seriously. And at the April 1923 meeting of the Royal Astronomical Society, after the confirming results of the 1922 eclipse were known, a satisfied Eddington declared about general relativity and the bending of starlight around the Sun: "I think it was Bellman in *The Hunting of the Shark* who laid down the rule 'When I say it three times, it is right.' The stars have now said it three times to three separate expeditions; and I am convinced their answer is right."

It was now clear that, near a massive object, space is non-Euclidean—it is curved. The question may then arise: What is the shape of the *entire universe*, not just the local neighborhood of a massive object such as a star? But here again, Einstein was way ahead of the crowd. Confident that his theory was right and space was non-Euclidean, Einstein had already begun to consider the shape and evolution of the entire universe two years before the big event of the 1919 eclipse. His work would lead him to the most controversial hypothesis of his life. In 1917, while manipulating his field equation, Einstein unwittingly opened a Pandora's box.

7. Reprinted in German in Michael Gruning, *A House for Albert Einstein*, Berlin: Verlag der Nation, 1990, pp. 388–9.

CHAPTER 11

Cosmological Considerations

"I've often wondered how Einstein could ever bring himself to make such a simple assumption . . . the universe is so simple that we can analyze it in a one-dimensional differential equation—everything a function of time alone. Of course, Einstein had brilliant intuition, and he surely was awfully close to the truth—that's the way the universe looks." —James Peebles, Princeton cosmologist, 1990[1]

In February, 1917, Albert Einstein submitted a paper to the Prussian Academy of Science that marked the birth of modern cosmology. Here, Einstein applied the full power of his recently completed theory of general relativity to addressing questions about the universe as a whole. The paper was titled *Cosmological Considerations on the General Theory of Relativity*. Researchers have hypothesized that Einstein extended his quest to the entire universe because of an idea of Ernst Mach.[2]

Mach held that inertial forces observed in the world are due to the total system of the fixed stars as a reference frame. This is called Mach's law of inertia. It says that the total inertia of a mass point is an effect that is due to the presence of all other masses in the universe. To understand Mach's law of inertia,

1. In A. Lightman and R. Brawer, *Origins: The Lives and Worlds of Modern Cosmologists*, Cambridge, MA: Harvard University Press, 1990.

2. See for example, Max Born, *Einstein's Theory of Relativity*, New York: Dover, 1965, p. 362.

consider Foucault's pendulum. This pendulum, which hangs from a high ceiling and regularly swings across a large circle on the ground, is displayed in many museums of science and other public areas. Its principle was discovered by Jean Foucault (1819–1868).

As the pendulum swings back and forth over many hours of the day, one can clearly discern a shifting in the angle of the swing through time. What happens here is that the pendulum *stays true* not to the earth but rather to its own constant direction of swing, and the earth, in its rotation, is changing *its* tilt right under the pendulum. Mach considered that Foucault's pendulum keeps its inertia with respect to the fixed stars in the universe, independently of the earth.

Einstein's paper begins with a new analysis of the problem of an old equation, one due to Newton and to the French mathematician Simeon-Denis Poisson (1781–1840). Newton came to the conclusion that a finite universe could not exist. As early as the 1690s, Newton knew that since gravity pulls all massive objects toward all other massive objects, a static, finite universe was impossible. Why? One way to see this is to use the fact from physics that mass can be viewed as acting from its center—the center of mass. If you imagine a very large sphere of finite space filled with galaxies and stars and inter-galactic matter, you can view the force as concentrated at the center of all the mass in the sphere. All the matter in such a static, finite universe would be drawn toward the center of mass, producing an inward velocity by all the objects in the universe, and eventually a collapse of everything to that one point.

Newton then argued that if the universe had infinitely many stars, distributed over infinite space, this would not happen since there would be no center of gravity into which everything could fall. However, the argument was flawed because every point in an infinite universe could be viewed as the center of the universe, as in every direction one looks there are infinitely many stars. Later it was discovered that a mathematical limit argument was the way to approach this problem: assume a finite universe, then add stars radially in all directions ad infinitum. When such an argument was pursued, it became clear that even an infinite universe—if static with gravity the only long-range force—would eventually collapse upon itself.

In his paper, however, Einstein began by discussing the Newtonian idea of gravity, and stated the Poisson equation, a differential equation that relates the distribution of matter with changes in the gravitational field, ϕ. Einstein noted that at spatial infinity, the gravitational field, ϕ, tends toward some fixed, finite number. He noted that if we wanted to look at the universe as infinite in its spatial extent, then some limiting conditions must be imposed on the equations of his general theory of relativity. Einstein then argued that the correct condition to impose on the equations to get a limiting value for the gravitational field at infinity was that the average density of matter in the universe, denoted by ρ, must decrease toward zero more rapidly than $1/r^2$, where r is the distance from the center of this spherical universe tending outwards to infinity. This condition imposed a form of finiteness on the universe, Einstein said, although the total mass could be infinite.[3]

Einstein continued to ponder the model he was building for the universe, with the gravitational field of Newton and Poisson, ρ, replaced in Einstein's field equation of gravitation by the Rie-

3. Albert Einstein, "Cosmological Considerations on the General Theory of Relativity," reprinted in Albert Einstein, *The Principle of Relativity*, New York: Dover, 1923, p. 178.

mann metric tensor, $g_{\mu\nu}$, and the Newton-Poisson density of matter, ρ, replaced by Einstein's energy-momentum tensor $T_{\mu\nu}$.[4] These tensor quantities were the elements of the field equation of gravitation Einstein derived two years earlier as the final product of his theory of general relativity:

$$R_{\mu\nu} - 1/2\, g_{\mu\nu} R = -\kappa T_{\mu\nu}$$

(Where $\kappa = 8\ \pi\ G$, the equation being written more compactly in the form above.)

The question in his mind was how to generalize the Newton-Poisson relationship consistently in his tensor-valued field equation of gravitation, so that general relativity would apply in a meaningful way to the vast universe as a whole, not just the local neighborhood of a star or galaxy.

Following the assumptions he outlined about the average density of matter decreasing to zero faster than one over the squared radius of the universe, Einstein realized that his equation would have to satisfy an interesting condition: radiation emitted by the heavenly bodies will, in part, leave the Newtonian system of the universe, escaping out of the universe to be lost in the vastness of infinity. The fact that the gravitational field must become constant at spatial infinity told Einstein that in the same way that a ray of light would leave the universe and continue outward to infinity, so could a massive body such as a star. A star could thus overcome the Newtonian forces of attraction and "reach spatial infinity." He wrote: "By statistical mechanics this case must occur from time to time, as long as the total energy of the stellar sys-

4. Recall that energy and mass are equivalent by Einstein's famous formula $E=mc^2$.

tem—transferred to one single star—is great enough to send that star on its journey to infinity, whence it never can return."[5]

At this point Einstein made a stunning realization: the universe itself must be *expanding*—the stars, matter, and radiation all had to be flying outwards toward "infinity," or else the whole universe would collapse upon itself, whether it had a finite number of stars and a finite amount of matter or not.[6] Thus Einstein—from his own field equation—discovered the expansion of the universe. But he didn't believe his own conclusion. In his paper, he repeated that the stars have relatively small observed velocities (that is, none have been seen to fly off toward infinity, or else their velocities would have to be higher). He wrote: "We might try to avoid this peculiar difficulty by assuming a very high value for the limiting potential at infinity. That would be a possible way, if the value of the gravitational potential were not itself necessarily conditioned by the heavenly bodies. The truth is that we are compelled to regard the occurrence of any great differences of potential of the gravitational field as contradicting the facts. These differences must really be of so low an order

5. Ibid., p. 178.

6. The idea of infinity as applied to the *space* available to our universe can be viewed this way. Imagine a plane stretching as far as the eye can see in every direction. Now at the horizon, in any direction you look, the plane curves upwards and continues to rise with an increasing slope. Way up, in the *limit*, everything curving up all around you is joined at that one point very far above your head—the point at infinity. (This mathematical model is called a "one point compactification of the plane.") If you think of the space on which you stand as three-dimensional rather than two, or even four-dimensional to account for space and time, the above model works as an infinite-space universe.

of magnitude that the stellar velocities generated by them do not exceed the velocities actually observed."[7] If Einstein only *knew* then what we know today. The farthest galaxies we have observed are flying away from our direction (toward *infinity*) at velocities of over 95% of the speed of light.

But Einstein didn't know these facts. In his universe there was only one galaxy—the Milky Way. Even Andromeda, our neighbor at 2.2 million light years away, had by 1917 not yet been determined to be another galaxy. It was thought to be a nebula —a hazy patch of sky thought to be made of gas and dust— residing within our own galaxy, the universe. And here, in the Milky Way, stars do not move very fast. So Einstein did what seemed right to him—he ignored what his theory told him, and sought to change the theory to suit the reality he saw: a static universe that somehow doesn't fall inwards into its center.

Einstein remarked that someone else had already thought about the problem of applying general relativity to the problems of the cosmos, but had chosen an approach tantamount to giving up—de Sitter, who had addressed these problems in a paper published in November 1916 by the Academy of Sciences of Amsterdam. However, Einstein continued: "I must confess that such a complete resignation in this fundamental question is for me a difficult thing. I should not make up my mind to it until every effort to make headway toward a satisfactory view had proved to be in vain."

Einstein's field equation of gravitation is mathematically elegant, which is why Einstein held on to the belief that his equation was correct even before experimental evidence for general

7. Ibid.

relativity was obtained, and this is why he later said he would have felt "sorry for the Lord" if the experiments had failed, as "the theory is correct." Thus faced with the apparent problem of reconciling a seemingly static universe with the beautiful equation that implied an expanding universe was a serious trauma. But, as he said, he had to attempt an accommodation of reality with his equation until every effort had proved to be in vain. And so he did. Einstein *changed* his perfect equation, which had served him and physics so well in describing natural phenomena. The equation:

$$R_{\mu\nu} - 1/2\, g_{\mu\nu}R = -\kappa\, T_{\mu\nu}$$

was changed to:

$$R_{\mu\nu} - 1/2\, g_{\mu\nu}R - \lambda g_{\mu\nu} = -\kappa T_{\mu\nu}$$

Here, he added a simple constant, denoted by the Greek letter Lambda, λ, multiplying his metric tensor $g_{\mu\nu}$. The modification was done in a careful way aimed at preserving the important physical characteristics that must be captured by a meaningful equation. The alteration Einstein made in the equation was designed to have little effect on local phenomena such as the motions of planets, but great effect at huge distances. It was an ingenious plan—something only Einstein himself could do.

Einstein made use of the fact that his space was non-Euclidean. The curvature of the space Einstein considered was expressed by ten quantities, which corresponded to the ten metric-tensor coefficients g within the four dimensions of spacetime. Einstein tweaked each one of the elements of his metric tensor by a small amount, λ. What he did was to manipulate cleverly the geometry of the universe to make it fit the equation. The term, λ, which enabled him to do this, would later be called the *cosmological constant*. And Einstein would never be able to live down its creation. The cosmological constant would haunt him for the rest of his life.

Einstein's equation containing the cosmological constant had many desirable properties. His equation was the very first mathematical model of the universe as a whole. In this model, the universe is static—neither expanding nor contracting. It is spherical in shape and bounded. It has constant curvature. And the problem of the Newtonian infinite is solved since the universe is *finite*, and yet without boundaries. To see how a model can be finite and yet unbounded, consider the two-dimensional example of the surface of a sphere. Here a great circle is the shortest distance between two points. If such a curve is followed on the surface of the earth, eventually one reaches the point of departure after having circled the globe. But there are no boundaries on such a curve (called a geodesic) on the surface of the earth, and yet the surface is finite. Einstein's universe is a three-dimensional analogy of the surface of the earth. Here, a ray of light or a particle traveling along a geodesic (a curve of shortest distance between two points) will eventually return to its point of departure—however, this will take a very long time. Such a universe is finite but unbounded. Einstein's universe has curvature that is time-independent. The universe is homogeneous, that is, it looks the same everywhere. It is also *isotropic*, that is, it looks the same in every direction the observer may look—there is no preferred direction in space.

The radius of curvature of Einstein's three-dimensional spherical universe is strongly related to the cosmological constant, λ, and both the radius and λ depend on the total amount of matter in the entire universe. If the universe has a lot of mass, its radius of curvature is smaller—the giant sphere is more curved when it contains more mass. But as the matter is dilated, the curvature of space diminishes. Einstein considered these properties realistic. He was also concerned with the *density* of the mass in the universe, that is, the ratio of mass to space: how is the total mass of the universe distributed within the entire huge sphere of space?

The average density of mass in Einstein's universe is assumed to be constant everywhere in space. Einstein used the local results of his general theory of relativity in arguing this point. "According to the general theory of relativity," he wrote in his article on cosmology, "the metrical character (curvature) of the four-dimensional space-time continuum is defined at every point by the matter at that point and the state of that matter." Since matter varies widely in space, he argued, the metrical properties of space are necessarily extremely complicated.

But Einstein offered a way out. He said that if we are concerned with the large-scale structure of the universe, rather than local properties such as the high curvature of space around massive objects, then the *average* density of space should be the pertinent parameter. Thus, "we may represent matter to ourselves as being uniformly distributed over enormous space, so that its density of distribution is a variable function which varies extremely slowly."

Einstein had introduced the concept captured by a parameter ρ, the average density of matter in the universe. This concept of the average density of matter in the universe would remain an important staple of all cosmological theories through the end of the twentieth century. Einstein thought he had found another desirable property for his universe, for the model he derived as a solution to his equation implied that when the cosmological constant was not zero, the equation was not satisfied when the mass density parameter, ρ, was equal to zero. So the equation did not apply, he thought, in a universe containing no matter. This property he found reassuring.

At first, the introduction of the cosmological constant into Einstein's equation seemed innocuous enough. It was Einstein's own great equation—it would seem he'd have every right to change it if he pleased. But very soon the first challenge arrived. Back in Holland, old Willem de Sitter with his mane of white hair and goatee was still hard at work on cosmology and general relativity. He was the good disciple, along with his friends Ehrenfest and Lorentz, keeping up their work of embellishing Einstein's creation. Earlier in that same year, 1917, in which Einstein's paper on cosmology appeared, de Sitter had already published his own paper, causing Einstein his greatest embarrassment. For de Sitter's paper presented *another solution* to Einstein's field equation cum cosmological constant. De Sitter's solution allowed for a universe with no matter at all—a universe of empty space.

Einstein was embarrassed because he had been guided in his quest to explain the cosmos by Mach's idea that the mass distribution of the universe should set inertial frames (Foucault's pendulum swinging as it does in response to the combined force exerted on it by all the mass in the universe). Einstein seems to have believed in Mach's principle, and during his Prague year had written about the plausibility of the conjecture that the total inertia of a mass point was due to the presence of all other masses—a sort of interaction of a point mass with all other masses in the universe. Later in Zürich he was as convinced as ever that the principle was valid, and even wrote to Mach that if the bending of light should be discovered, it would confirm Mach's hypothesis.

But when he wrote his 1917 paper on cosmology, dealing in

part with the Mach hypothesis, he had some trepidations—for he wrote to his friend Paul Ehrenfest that he had "again perpetrated something about gravitation that exposes me to the danger of being confined to a madhouse." In a conversation with de Sitter before the paper was published, Einstein mentioned the possibility that inertia originates strictly because of the existence of all the matter in the universe. But now that Einstein's cosmological paper was published, de Sitter's solution implied that mass wasn't necessary in the universe to set such inertial frames. And de Sitter's solution to Einstein's equation had another very important characteristic—one that was overlooked at the time de Sitter's paper appeared. The universe according to de Sitter's solution could well exist with no matter at all. However, if matter existed in the universe—then the universe was *not static.*

There was some kind of cosmic repulsion force acting on all the masses, for they were flying away from each other. This phenomenon would manifest itself, according to the solution to the equation, only at large distances. The astute de Sitter, therefore, began to search for astronomical reports of a cosmic expansion, but he found none. The expansion in de Sitter's universe, it should be noted, is not a simple one. It would be consistent with the inflationary universe theory developed in 1980 by the cosmologist Alan Guth of MIT. But even bigger blows to Einstein's theory incorporating the mischievous cosmological constant were just around the corner.

While Einstein never said so himself, he was undoubtedly convinced that the correct field equation of the universe should not have a solution that does not include matter. An empty universe would be unappealing for many reasons beside its violation of Mach's idea about inertial frames. After de Sitter's paper appeared, Einstein spent much effort throughout the next two years trying to find an error in de Sitter's solution to Einstein's cosmological equation. But Einstein failed. In 1919, he tried something else. Einstein took his equation with the cosmological constant and added the assumption that the energy-momentum tensor T was due to electromagnetism. The assumption was driven by his hypothesis that electrically-charged particles were held together by gravitation.

This approach was Einstein's first attempt at *unifying* the theories of physics: here, in a way, he was trying to unify electromagnetism and gravitation. The quest for a unified field theory would occupy most of Einstein's time for the rest of his life, but he would not succeed in finding an equation to describe all the laws of physics. At any rate, in the years following 1917, Einstein stopped talking about Mach's idea of inertia and the masses of the universe. And he began to lose faith in the cosmological constant. Soon would come the coup de grace—at least from Einstein's point of view.

The blow to the cosmological constant was dealt by the works of two American astronomers. The same year that saw the publication of the papers by Einstein and de Sitter, 1917, also saw the publication of a paper by Vesto M. Slipher, an astronomer at the Lowell Observatory in Flagstaff, Arizona. Slipher looked through his telescope at spiral nebulae. At that time, these spirals were considered to be part of the Milky Way. But Slipher noticed that in addition to a rotational velocity of stars within these spiral nebulae, there was also a large, consistent shift of the spectral lines toward the red end of the spectrum. When the redshift was translated into a velocity using calculations of the Doppler effect, the result was that the spiral nebulae were receding from us at very high speeds—some of them over two million miles an hour. What Slipher had in front of him, but could not know it at that time, was evidence for the expansion of the universe. He could tell that most of the nebulae he was observing were moving away from us very fast (except for the few we know today are nearby galaxies), but he did not know these were separate galaxies, nor did he know the distances to these galaxies. But soon the picture would become much clearer through the work of Edwin Hubble.

In order for Hubble to make his stunning discovery about our universe, an astronomer at the Harvard College Observatory had to invent a breakthrough astronomical technique. Henrietta Leavitt (1868–1921), of Harvard College Observatory, worked on cataloging variable stars observed at Harvard's southern observatory in Peru. Leavitt was studying the light curves of variable stars in the Large and Small Magellanic Clouds. These are two satellite galaxies of the Milky Way, although they were not yet known to be separate galaxies. The two Magellanic Clouds were discovered by the crew of Magellan's ship during his voyage around the world in 1686. They appeared as hazy bright patches in the night sky near the south celestial pole. Since each Magellanic Cloud looked like a cohesive patch of stars, it seemed reasonable that the distances from us to the stars in a patch should be about the same (on an astronomical scale).

This turned out to be a good guess, as it allowed Leavitt to make an amazing discovery. She found a direct relationship between a variable star's apparent magnitude and the star's period, or cycle of changes in magnitude. Since the stars in a Magellanic Cloud were approximately at the same distance from Earth, the relationship between apparent magnitude and period could be assumed to exist also between the absolute magnitude (that is, the magnitude for a standard, pre-specified distance) and period of variability. The stars Henrietta Leavitt was studying were of a particular kind: Cepheid variables. They are named after the first star discovered to have a very regular period of brightening and dimming: the star delta in the constellation Cepheus.

By 1912, Leavitt had determined the relationship between magnitude and period for twenty-five stars. She continued her hard work of comparing each star's variation in brightness with the star's period, and found the exact mathematical relationship that existed between the two variables. The brighter the Cepheid variable appeared, the longer its period. From this relationship between the two astronomical variables, distances to stars could be determined—wherever members of this specific type of star were found. By 1917, Leavitt's technique could be used for determining the distance to any Cepheid variable—even those in faraway galaxies. Leavitt had just given astronomy its first "standard candles," as they are now known, for determining cosmic distances. In 1998, Saul Perlmutter and his team would report on their use of another type of standard candle—Type Ia supernovae—to accomplish a similar task: determining the distances to much, much farther galaxies. For such distant galaxies the Cepheid variables technique couldn't work because these stars are too far away, while the very powerful supernova explosions could be discerned by careful photography and very powerful telescopes. At any rate, it was time for Hubble to discover the first consistent evidence for cosmic expansion.

Edwin Hubble (1889–1953), after whom the Hubble Space Telescope is named, is considered the greatest astronomer of the twentieth century. He was born in Marshfield, Missouri, on November 20, 1889. Hubble studied law at Oxford University, but soon turned to astronomy, and enrolled as a graduate student in astronomy at the Yerkes Observatory of the University of Chicago. Then he was called to serve in the war. Hubble was 30 when he was discharged from service in the American Expeditionary Force in Europe in 1919, at the end of World War I, and joined the staff of astronomers at the Mount Wilson Observatory in California. At that time, the observatory had the world's largest telescope: the 100-inch Hooker reflector. In 1923, Hubble began an observation program of searching for novae in the Andromeda Nebula—the largest spiral nebula in the sky (which we now know as the nearest galaxy, other than the two Magellanic Clouds). When Hubble looked more carefully at a photograph of the first nova he thought he had discovered in the Andromeda Nebula, he realized that what he had was actually a Cepheid variable.

By that time, the astronomer Harlow Shapley (1885–1972), working at the Mount Wilson Observatory, had estimated the distance to the Magellanic Clouds using Henrietta Leavitt's method. Thus, when Hubble realized that the star he had photographed in Andromeda was a Cepheid, he got very excited. After studying the light curve of the star, Hubble was able to estimate the distance to Andromeda as 900,000 light years (although today we know it to be larger—2.2 million light years). This finding was enough to establish that Andromeda was a separate galaxy, not part of the Milky Way. This discovery settled what by then had become known as the Great Debate in astronomy: Do island universes exist, or does the universe consist only of the Milky Way, with everything seen in the sky included in it?

Having established the existence of one galaxy completely separate from the Milky Way, Hubble trained the giant 100-inch telescope at other nebulous regions in the sky to try to determine whether these, too, were separate galaxies. Over the next few years, Hubble devoted all his energies, working alone and with his colleague Milton Humason (1891–1957), to observing galaxies with the 100-inch and later 200-inch telescopes. By 1929, Hubble had analyzed both the distance and the Doppler redshift effect for two dozen galaxies.[8] Here, Hubble made his great discovery: in general, galaxies are receding from us at a velocity that is proportional to their distance from us. (The redshift allowed him to compute a recession velocity, while the distance was estimated by Leavitt's rule for Cepheids). There was a clear, straight-line relationship between recession velocity and distance. The slope of the line is now called Hubble's Constant, and the straight-line law of recession velocity to distance is called Hubble's Law. The only logical explanation for Hubble's Law is that the universe as a whole is expanding: like a rising raisin cake.

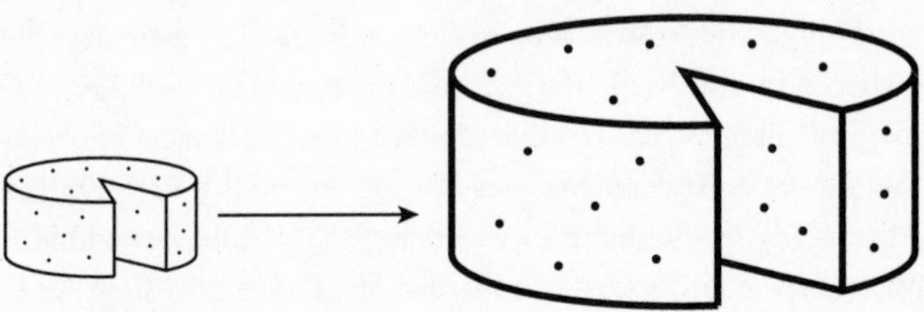

Hubble's Law changed our understanding of the universe. A static model was no longer appropriate. And, for a universe expanding at a constant rate—as seemed to be the case from Hubble's data of relatively close galaxies—the cosmological constant in Einstein's equation would not be necessary. In 1931, after visiting California and seeing the astronomers' calculations, Einstein conceded that the cosmological constant in his equation was unsuitable and he formally abandoned it. But even

8. Most galaxies were found to exhibit a redshift—indicating that they are *receding* from us. Some relatively close galaxies exhibit a blue-shift, meaning that they are getting closer to us. This phenomenon, however, is by far an exception to the general rule, and happens because a nearby galaxy is responding to the gravitational pull of the Milky Way, or is just a random motion in our direction.

before the announcement of Hubble's discovery of the expansion of the universe, Einstein became disappointed with his cosmological constant on purely theoretical grounds. His distaste for the constant had to do with two other models of the universe based on his own equation with the cosmological constant.

Alexander Friedmann (1888–1925) was born in St. Petersburg and studied meteorology and mathematics. He became interested in general relativity while working at the Russian Academy of Sciences. In studying Einstein's equations, Friedmann decided to abandon the assumption of a static universe, while keeping Einstein's assumptions of homogeneity and isotropy. Friedmann realized while solving the equations that the cosmological constant was not needed for this universe. Einstein thought that Friedmann's model was wrong—that Friedmann had made an error in his solution—and wrote about it. But later Einstein realized that it was he who had made an error in his objection to Friedmann's solution—and he retracted his complaint, calling Friedmann's work "clarifying." This was Einstein's second theoretical shock brought about by the cosmological constant. Then there was the work of a Belgian priest and mathematician named Georges Lemaître. In 1927, Lemaître learned about Slipher's first observations of redshifts and proposed a mathematical model for an expanding universe.

In 1923, Hermann Weyl (1885–1955) and Eddington studied what happens to particles subjected to de Sitter's model of the universe, and found that these were receding from one another. In a letter to Weyl following this result, Einstein wrote: "If there is no quasi-static world, then away with the cosmological constant!"[9]

9. Reported in Abraham Pais, *'Subtle is the Lord . . . '*, New York: Oxford University Press, 1982, p. 288.

CHAPTER 12

The Expansion of Space

"The universe is the ultimate free lunch."
—Alan Guth

In a seminal monograph entitled *The Cosmological Constant Problem,* Nobel Prize winning physicist Steven Weinberg, one of the few people who understands the minutest nuances of Einstein's equations of general relativity, wrote: "Unfortunately, it was not so easy simply to drop the cosmological constant, because anything that contributes to the energy density of the vacuum acts just like a cosmological constant."[1] In introducing his cosmological constant, Einstein had created a new mathematical tool for science—and even he could not take it away from the world. Whether this device was useful for physicists and cosmologists in explaining theories of the universe was now an important question.

A positive cosmological constant can act as a repulsive force to counteract the attractive force of gravitation. When Einstein first introduced his constant, he wanted something to push out-

1. Steven Weinberg, *The Cosmological Constant Problem*, Morris Loeb Lectures in Physics, Harvard University, 1988.

wards a universe that, according to the field equation he devised, wanted to collapse. This was a kind of artificial force that Einstein imposed to stop the universe from falling inwards due to gravitation. So when it was found that the universe was expanding, Einstein dropped the cosmological constant.

When Alexander Friedmann threw away Einstein's assumption of a static universe and solved Einstein's original field equation, obtaining the expansion of the universe, a new chapter in modern cosmology was opened. This led the way for Lemaître and others to ask the obvious question: If the universe is expanding, then when did that expansion begin? The intuitive answer was that at some time in the distant past, the universe was very, very close together—and therefore extremely hot and dense. From that very dense small clump of matter and energy, the giant universe somehow evolved by rapid expansion. In a radio talk on the BBC in the late 1940s, Cambridge cosmologist Fred Hoyle coined the term "big bang" to describe the gargantuan explosion that started our universe and its expansion.

The curvature of space-time increases with the mass of the object, so when the entire universe was condensed into a very small place, the curvature of space was extremely high. When the entire universe was condensed to a point, time stood still because at the point—the spacetime *singularity*—the mass density was infinite, and the equations of time and space no longer applied. Time could not be defined at the singularity. This idea prompted Lemaître to describe the beginning of the universe as "A day without a yesterday."

In 1965, Roger Penrose of Oxford University wrote a paper in which he used topological ideas to describe how a very massive object can collapse to a point—virtually get crushed under its own weight. When this happens, the outcome is a black hole, and the original understanding we have of this process from a relativistic point of view comes from the pioneering work of Karl Schwarzschild, who was first to solve Einstein's field equation, finding what we call today the Schwarzschild radius of a star. This is the radius such that if a star's actual radius falls below that number (determined by the star's mass), the star will collapse to a point. The Schwarzschild radius is the "point of no return"—any object or light ray that falls into a black hole beyond the invisible Schwarzschild radius of a black hole will be lost forever. Penrose proved that there lies a point like no other at the very center of a black hole. This point is a spacetime singularity. Here, the curvature is infinite, and time ceases to exist.

By Penrose's ingenious argument, the collapse of the mass of the star reaches a rate from which there is no escape. Mathematically, a *trapped surface* develops inside the star, and as the accelerating collapse continues, nothing can stop it and the result is the singularity: a point where mathematics and physics as we know them no longer apply. Even if the star deviates in shape from perfect spherical symmetry, the collapse to a singularity still occurs.[2] Incidentally, long before the singularity, at the Schwarzschild radius, time stops—as observed by someone from outside. Thus a person falling into the black hole will look to an observer to be freezing on the surface of the Schwarzschild radius. The falling person will never be aware of this state of being.[3]

The problem with a singularity is that we can't really understand it within mathematics or physics as we know them. In

2. Roger Penrose, "Gravitational Collapse and Space-Time Singularities," *Physical Review Letters*, 18 January, 1965, pp. 57–59.

3. For more on this fascinating phenomenon, read Leonard Susskind's elucidating and entertaining article, "Black Holes and the Information Paradox," *Scientific American*, April 1997, pp.52–57.

mathematics, a singularity is a point where something pathological happens. An easy example demonstrates the idea of a singularity. If you think of a *function* of a variable, $y=f(x)$, the function could be well-behaved: smooth and continuous, and with a well-defined derivative (the instantaneous *rate of change* of the function). But now think of a function such as $y=1/x$. When $x=1$, $y=1$; when $x=2$, $y=1/2$; when $x=1/2$, $y=2$; when $x=-2$, $y=-1/2$. But what happens when $x=0$? Here the function is not defined. We've just left the realm of the reasonable, the rational, the sane. Is its value *infinity*? But then it is just as logical that it should be minus infinity. And even if it were defined as infinity, what is its rate of change? The derivative here no longer exists—although it exists at any small neighborhood of the point $x=0$.

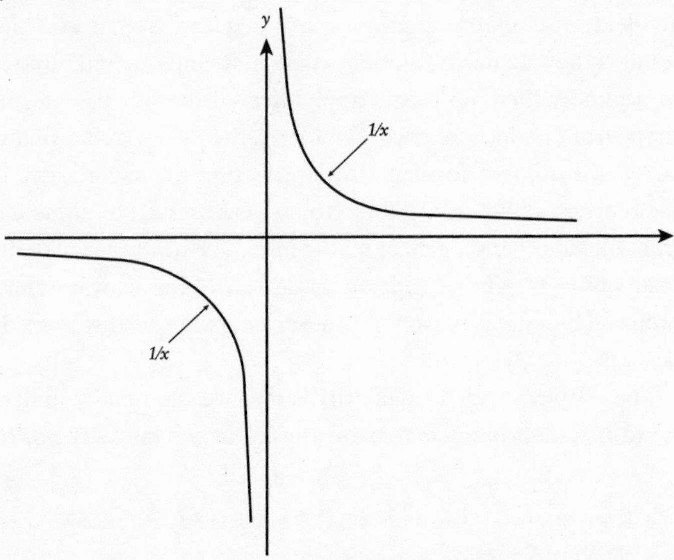

At the heart of a black hole (and at the beginning point of the universe) all laws break down in the same way as the function $y=1/x$ breaks down at $x=0$. Gravity is infinite, the curvature of space-time is infinite, and time stops.

Penrose's argument was reversed—as is allowed under the laws of general relativity—from a collapse to an expansion. Thus if we reverse the movie from a collapse, we see the universe originating from a *white hole*, as it is called—a spacetime singularity. Since mathematics and physics don't work at the singularity—and we have no answer to what happened *before* the explosion, if that is when space and *time* began—theories of the beginning of the universe typically start with what happened a tiny fraction of a second after the big bang. Here, there are two main approaches.

The standard big bang theory states that after the tremendous explosion, space began to expand. Some time later baryonic matter was created: matter as we know it, protons and electrons and neutrons, all in a state of a very hot primordial soup that slowly began to cool as it expanded. The early universe was opaque because it was so dense: radiation in the form of photons was constantly being absorbed and re-emitted. Only when the universe was about 300,000 years old did it become transparent enough so that photons could travel in straight lines. This opacity of the early universe acts as a boundary.

As we look farther and farther into space with increasingly more powerful telescopes, we also look further and further back in *time*. Eventually, we reach a limit. If our telescopes could actually see as far as 14 billion light years away, we would not see anything at all at that distance. By most estimates the universe is between 12 and 14 billion years old, so 14 billion years is close to the big bang—closer to it than 300,000 years. But at an age of 300,000 years, the universe was opaque—so we cannot see anything at that distance or farther.

A billion years or so after the big bang, the first stars and

Alan Guth
Donna Coveney/MIT

galaxies began to form. Clusters and superclusters of galaxies emerged over time. The universe continued its expansion, eventually reaching its current size.

The *inflationary universe theory* is an alternative to the standard big bang theory of the development of the universe. Alan Guth, the Weisskopf Professor of Physics at MIT, invented the powerful inflationary universe model in 1979. Alan Guth was born in New Brunswick, N.J., in 1947. He received his B.S. and M.S. degrees from M.I.T. in 1969, and continued to study there for a doctorate in physics. He received his Ph.D. from M.I.T. in 1972. He went on to a position of instructor of physics at Princeton, and a research associate at Columbia. Then Guth worked in particle physics as a postdoc at Cornell University. In the fall of 1979, he took a year off from his job at Cornell to work at the Stanford Linear Accelerator. There, in December, he had a brilliant idea. Guth's theory tries to explain what happened in the first fraction of a second after the big bang in a way that offers solutions to two big puzzles in cosmology: the flatness problem and the horizon problem.

The flatness problem arises when one espouses a particular belief about the universe. This belief is that the geometry of the universe is "flat," or Euclidean. Guth and his cosmological predecessors reached this conclusion based on estimates of the density of matter in the universe. They computed a *critical* density for the universe: a density that, if achieved, would make the universe neither collapse nor continue to expand at the same or greater speed, but instead expand at a declining rate, eventually approaching zero. These scientists estimated that the present density of the universe is close to the critical density (where close means anywhere from a fraction of the critical density to twice the critical density). The scientists reached their conclusion by extrapolating estimates back to a time when the universe was very young. They found that when the universe was at an age of one second after the big bang, its density agreed with the critical density to 15 decimal places. This led them to believe that the universe is incredibly flat. But why? The standard big bang theory could not explain this conclusion.

The second problem with the standard big bang theory is the horizon problem. The horizon, as on Earth, is the point beyond which we cannot see. Within the context of relativity, if a light signal is sent to us from a place farther away from us than light would have had time to travel, that place is beyond our horizon.

In April, 1998, Esther Hu of the University of Hawaii and her colleagues reported seeing the most distant galaxy using the world's largest telescope, one of the twin ten meter Keck telescopes. The faint galaxy is about 13 billion light years away. The universe is about 14 billion light years old. Now suppose

Esther Hu

that Hu, or another astronomer, were to look in the opposite direction in the sky and see another galaxy at a distance of about 13 billion light years from us. Clearly, these two galaxies are each *outside the horizon* of the other. Why? Because the universe is 14 billion years old, and for light to have a chance to travel from one of these two galaxies all the way to the other would take 13 + 13 = 26 billion years, close to twice the age of the universe.[4] There is no way for the light to arrive there. Furthermore, because of the expansion of the universe, these two galaxies are receding from each other at speeds comparable to

4. Here and elsewhere, numbers in billions of years are somewhat off due to the continuous expansion of space.

that of light, so the light from one galaxy may take forever to reach the other galaxy.

The horizon problem arose from studies of the cosmic background radiation. How can this radiation be so homogeneous (to within 1 part in 100,000) when coming from every possible direction in space? Since points along various directions in space could not "see" each other once they were at distances beyond their mutual horizons, there was no possible exchange of information among them to effect the homogeneity.

To explain these effects, Guth proposed the inflationary theory, which says that during the first fraction of a second, the universe expanded at a tremendous, exponential rate. Guth's ideas about cosmology invoked a mechanism well-known to particle physicists, and by which an unusual form of matter in the early universe could have created a gravitational repulsion providing the driving force of the expansion of the universe. This propelled expansion maintained the homogeneity of space and provided the desired solution to the horizon problem. At the same time, the force-driven expansion is seen as driving the universe toward the critical density.

Inflation has many other ramifications. A cosmological constant in Einstein's equations makes the inflation theory more believable. Perhaps the cosmologists who met to discuss the new findings about the accelerating expansion of the universe were drawn to the cosmological constant by their desire to preserve the flatness hypothesis and hence the inflationary theory.

Alan Guth was first led to the idea of the inflationary universe from his study of magnetic monopoles—particles with only one magnetic pole instead of the usual two poles a magnet has—which should exist in the universe according to some theories, but apparently do not. His work was guided by the theory of the Higgs field, a theoretical tool used in particle physics. Higgs fields have never been detected in practice, but some scientists think they are responsible for broken symmetries in nature. Guth's inflationary universe theory assumes that our universe is a part of a larger super-universe, and that it evolved as the result of a vacuum fluctuation in that larger universe. Guth has also proposed that other "baby universes" may exist, spawned at other locations within the mother universe. He's even suggested that a super-advanced civilization might be able to create such a baby universe in the laboratory.

How do we know that the big bang really happened? If galaxies are receding from each other, then in the past they were closer. Extending this principle all the way back in time, we get that at one point everything had to have been clumped together. But how do we know this really happened? If the universe began in a big bang and expanded immediately afterwards, continuing to do so to our time, then it had to have started very hot, and it must keep cooling continuously as it expands. In the 1950s, the theoreticians George Gamow, Ralph Alpher, and Robert Herman suggested that telltale radiation of the big bang should still be present in the universe. That is, since the universe keeps cooling down from the immense temperature of the big bang, it must now have reached a given temperature—one that scientists should be able to measure.

In the 1960s, Robert H. Dicke and James E. Peebles of Princeton University made a similar prediction, and actually worked out the energy of this radiation, the *blackbody radiation* due to the big bang. Blackbody radiation is the radiation emitted by all bodies that are at a temperature above absolute zero. Everything emits some radiation which can be detected. The clearest example of this radiation is infrared radiation emitted by warm bodies. But even colder bodies emit radiation, at lower energy levels.

When photons were freed from the primordial soup as the universe became transparent about 300,000 years after the big bang, they started moving in straight lines and have been traveling continuously since then. Because of the Doppler effect, these photons have been losing energy and their present level of energy, and thus their wavelength, can be computed theoretically.

In 1965, two radio astronomers working at Bell Laboratories discovered just what the theorists had predicted—without knowing about the prediction. The two, Arno Penzias and Robert Wilson, later received the Nobel Prize for discovering the background microwave radiation. In 1989, the Cosmic Background Explorer satellite (COBE) was launched by NASA to measure the radiation with more precision. The results of the tests for the radiation from all directions in space were remarkably uniform, and corresponded to a temperature of 2.7 degrees above absolute zero (2.7 Kelvin). This discovery, along with Hubble's law, is considered one of the most important astronomical findings, and supports the big bang theory. The uniformity of the radiation is believed to lend support to the inflationary universe theory. Stars and galaxies and galaxy clusters are all believed to have formed from the ripples in the energy of the early universe and the big bang.

Another proof we have for the big bang theory is the relative abundances of chemical elements in the universe. Scientists have calculated the expected prevalence of elements as predicted by the big bang theory. The level of energy unleashed by the big bang implies that the universe should contain roughly 75% hydrogen, and 25% helium. All the elements other than these two (and the isotopes deuterium, helium-3, and lithium-7) should account for only traces of the total mixture of compounds in the universe. These heavier elements—making up everything around us including everything we are made of (although hydrogen is a major component in our bodies as well)—were created later, by nuclear reactions inside stars. Continuing studies of the makeup of the elements in the universe have all confirmed this hypothesis. These studies provide compelling evidence for the big bang theory.

The big bang began the expansion of the universe. Whether or not the expansion started exponentially, as suggested by inflation, the question is: What is the nature of the cosmic expansion? The best analogy for the expansion of the universe is given in the book *Origins: The Lives and Worlds of Modern Cosmologists*.[5] Take a rubber band and mark points on it in ink, say, every quarter inch. Each one of the ink dots on the rubber band represents a galaxy in space. Now continuously pull the rubber band between your hands. Notice that as you pull, the dots get farther apart. The distance between every two neighboring points on the band (galaxies in space) increases. Two adjacent points are now more than one quarter inch apart. But what happens to non-neighboring points? These are stretched apart even faster—the distance between them grows at a greater rate. If you stretch the band enough so that two points originally separated by one quarter inch are now half an inch apart, then two points that are separated by a third one are now a full one inch apart. This is what happens in the cosmic expansion.

As space expands, relatively close galaxies move away from one another at a lower speed than those that are farther away. The rate of separation of any two galaxies is *proportional* to their distance. This is exactly Hubble's Law.

Notice another very important property. As you stretch the rubber band, every point separates from its neighbors in exactly

5. Alan Lightman and Roberta Brawer, *Origins: The Lives and Worlds of Modern Cosmologists*, Cambridge, MA: Harvard University Press, 1990, p. 8.

the same way. There is no favored location on the rubber band: every point can be viewed as the center from which all other points recede. The universe has no obvious center, nor any edges. Even if we see every galaxy recede from *us*, every galaxy experiences the same phenomenon—the same illusion of being the center of the expansion.

What causes the expansion? Space is being created, or stretched. Space, as we know from general relativity, is plastic. It is a flexible medium whose geometry can change with the effects of gravity. Space is not an emptiness, a void, as it may seem. Space constantly expands like a rising cake. The galaxy discovered by Esther Hu and her colleagues is receding from us at 95.6% of the speed of light. This is what happens—with respect to us on Earth—to the most distant galaxies we can see. Galaxies that are seven billion light years from us seem to recede at a speed of about half the speed of light. And as we get to galaxies that are closer, the speed of recession—with respect to us—is slower.

To understand this weird, un-intuitive phenomenon, it is best to assume that the universe is infinite. If the universe is infinite, then any point is its center, and an observer at any point would see galaxies receding with the same property that closer galaxies recede at relatively small speeds while farther ones at increasing speeds relative to their distances. In the cosmic raisin cake, every raisin "sees" the other raisins expanding away at speeds that are proportional to their distances. This is the idea of a *uniform* expansion of the medium we think of as empty space.

The new observations of faraway supernovae imply that space is not only expanding—it is accelerating its expansion. So something is pushing space outwards. What could it be? According to quantum physics, space, the "vacuum," is not a vacuum at all—it is teeming with energy. Virtual particles appear and dis-

Evolution of the Universe
Illustration ©Shigemi Numazawa

appear continuously in what we think of as empty space. There is a great amount of energy in what looks like perfect emptiness, and we don't understand this energy or where it's coming from. The vacuum is like a contracted spring that wants to burst out. The pressure exerted by the invisible spring packed with energy makes the space in which it is hidden expand. But the spring relaxes at a much slower rate than the expansion it is causing, and so the expansion is accelerating its pace. The energy of the vacuum, the force pushing space outwards, is modeled by Einstein's cosmological constant.[6]

6. If we denote the energy of the vacuum by P_V, then Einstein's cosmological constant is given by: $\lambda = 8\pi G\, P_V$.

CHAPTER 13

The Nature of Matter

Neutrinos, they are very small
They have no charge and have no mass
And do not interact at all
The earth is just a silly ball
To them, through which they simply pass
Like dust maids down a drafty hall

—John Updike, 1960.

One of the most important issues facing scientists in quest of the nature of the universe is the question of the constitution of matter. What is matter? Is the universe dominated by matter, or are there other elements that play a major role in the evolution and properties of the universe? In the context of general relativity, the question about matter determines the constitution of Einstein's energy-momentum tensor, *T*. Matter behaves differently within the frameworks of the two important theories of physics in the twentieth century: general relativity and the quantum theory. General relativity determines the large-scale properties of matter (and space and time), while the quantum theory determines the small-scale properties of matter. The former is a perfectly deterministic theory, while the latter is intrinsically probabilistic in nature: answers to questions within the quantum realm are given in terms of probability distributions, not exact numbers. The quantum theory, among its other great triumphs, has also led to the discovery of previously-unknown particles of matter.

When the quantum theory was first discovered early in the twentieth century, physicists knew only of neutrons, protons, and electrons. Then a particular form of radioactive decay was discovered, where a neutron ejects an electron and a proton. By comparing the relative amounts of energy present in a system before and after this reaction, Wolfgang Pauli hypothesized in 1930 that an unknown particle must also be released by the reaction. A year later the Italian-American physicist Enrico Fermi named the projected particle a neutrino ("little neutron" in Italian). The neutrino was believed to carry away from the radioactive decay exactly the missing amount of energy.

The neutrino was seen as a charge-less particle, and until June 1998, it was believed to have no mass. At least no neutrino mass had ever been measured. In 1956, Fred Reines and Clyde Cowan discovered neutrinos being emitted in a nuclear reactor on the Savannah River. In 1995, after Cowan's death, Reines was awarded the Nobel Prize for discovering the particle whose existence was predicted a quarter of a century earlier. Thus a particle whose existence was "created" by scientists in order to account for energy that was mysteriously missing from the end products of a nuclear reaction was actually found. The story of the neutrino shows how theory and mathematics can be used in advancing knowledge, and that the confidence that good theoreticians may have in their theories can pay off in the results of future experiments. But the story of the neutrino was only beginning.

At around that time, the emerging knowledge of the mechanism of nuclear fusion convinced scientists that this kind of nuclear reaction must fuel the stars. And if the fires inside a star are nuclear fusion, which releases tremendous amounts of energy, then neutrinos must be emitted by stars, including our Sun. Scientists believed that the tiny particles, with no charge and zero or close to zero mass, were constantly arriving on

Earth from the Sun, but because of their incredibly small size, they were passing through the earth as if it weren't there. How could scientists detect these particles coming from the Sun?

The only way understood to detect neutrinos from space was to place large vats of various liquids in deep places such as mines, where they would be protected from other sources of radiation, and to look for the very rare interactions of the neutrinos with the molecules of water or other fluids in the vats. These experiments were carried out in many locations on Earth. In 1965, the first neutrinos whose source was outside the earth were discovered by Fred Reines and his colleagues in a gold mine in South Africa. In the meantime, another form of neutrino, the muon neutrino, was detected as a byproduct of nuclear reactions carried out at Brookhaven National Laboratory. Another type of neutrino was suggested to exist from the discovery of the tau particle at the Stanford Linear Accelerator. Experiments looking for solar neutrinos in salt mine reservoirs all detected numbers of neutrinos that fell far short of the number predicted by theory, stupefying experts.

But here too the theory preceded the experimental results. In the late 1950s, theories were developed by physicists showing that the neutrino might have a stunning property: it could change its form. Physicists use the word oscillate to describe this phenomenon. Thus an electron neutrino can turn into a muon neutrino or a tau neutrino (each named for the heavier particle, electron, muon, or tau particle with which the neutrino is associated). While one kind is more easily detectable, the others are not. So scientists concluded that some of the neutrinos coming from the Sun might be changing their type and avoiding detection.

In the 1980s, huge detection pools were constructed in the United States (the Irvine-Michigan-Brookhaven project in a mine in Ohio) and in Japan at the Kamioka Neutrino Observatory, located in a zinc mine 30 miles north of Takayama in the Japan Alps. The latter was an underground reservoir of 12.5 million gallons of ultra-pure water, surrounded by powerful light detectors whose purpose was to detect single emissions of a light ray resulting from a collision of a neutrino with an atom in the water molecule. In 1987, both giant neutrino detection projects found neutrinos that resulted from a supernova explosion in space, located in the Large Magellanic Cloud and observed from the Southern Hemisphere. The neutrinos had traveled *through the earth* to reach both of these detection sites. These were the first neutrinos confirmed to have come from outside our solar system, and their detection heralded the beginning of neutrino astronomy.

On June 5, 1998, an astounding announcement was made at a press conference in Takayama, Japan. The American-Japanese team of 120 physicists working at the Kamioka Neutrino Observatory were able to determine experimentally that the elusive neutrino has a mass. This discovery has far-reaching consequences—it can have an impact on our understanding of the nature of matter, the creation of the universe, and its destiny. The joint Japan-U.S. team was able to reach its conclusion that the neutrinos has a mass by detecting experimentally that neutrino really do switch "flavors" from the electron to the muon or tau variety. According to the quantum theory, anything that can do this must be a particle with mass. The actual mass of the neutrino could not be ascertained. But the fact that the neutrino had a mass meant that some of the "missing mass" of the universe had just been found. What is the "missing mass"?

Stars are clustered together in galaxies. A galaxy holds its stars together by the mutual gravitational pull of the aggregate of stars it contains. As we look deeper and deeper into space, we discover that the galaxies are not randomly located in space as we might expect. Rather, there is a structure. Benoit Mandelbrot of IBM Research Center and Yale University showed some years ago that the structure of galaxies resembles a fractal—an intricately ordered array, which is decidedly non-random, even if it looks so from a local vantage point. Galaxies form clusters of galaxies, which, themselves are arranged in super-clusters—and so on to larger and larger scales. Between the clusters are great bubbles of empty space, voids of magnitudes ranging in millions of light years.

In the 1930s, astronomers began to notice the clumping of the universe, the fact that the galaxies were arranged in clusters. Over the years, information was gathered about the distribution of matter over larger and larger horizons. In 1986, Margaret Geller, John Huchra, and Valerie de Lapparent of the Harvard-Smithsonian Center for Astrophysics constructed a map showing 6,000 galaxies in a slice of the northern hemisphere sky. The chart is centered on Earth, at the vertex of the "pie," and it extends to a distance of 650 million light years. The non-uniformity of the structure is clear from the picture, and we can even see that a fractal nature might apply. Where did this structure originate?

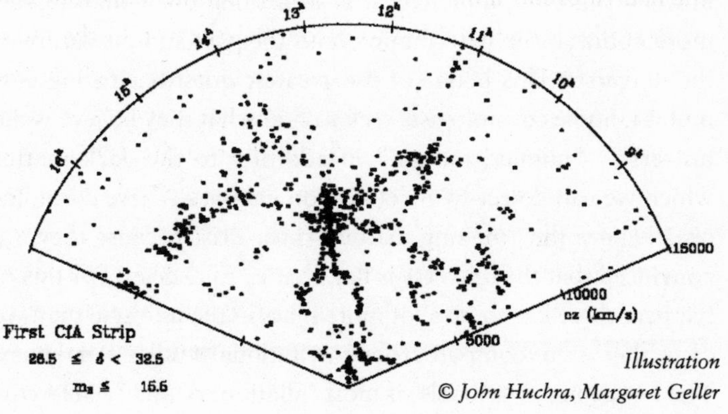

Illustration
© John Huchra, Margaret Geller

Quantum fluctuations in the first instants of time as the universe was created are believed to have formed bubbles of matter, which then became enlarged as the universe expanded. The mutual attraction of matter in the universe, due to the gravitational force, may have been responsible for the formation of the clusters and walls of galaxies we now observe. But when scientists tried to explain the gravitational effects within galaxies—the forces that keep galaxies together, they were stymied by a mysterious inconsistency.

In every galaxy astrophysicists studied, there was far less mass attributable to visible matter (stars or gas and dust) than the calculations had indicated should be present in order for the galaxy to hold itself together by gravity. The conclusion scientists could not escape was that the galaxies were permeated with an additional mass, accounting for 90% of all the mass in a galaxy. This mysterious, invisible yet perceivable mass was called "dark matter." This matter must be of a form unknown to science. It is not baryonic—atoms or subatomic particles—it is something never seen before. One of the biggest mysteries of astronomy is the nature of dark matter.

The properties of matter hold the key to cosmology. Is our universe matter-dominated, or is something other than matter more important in determining both the past and the future of the universe? This is one of the greatest questions facing cosmology. Some cosmologists seek to find what they believe is the universe's "missing matter," in addition to the dark matter which we can detect by its effects on galaxies. These cosmologists believe the "missing matter" must exist because they are convinced that the universe is flat, that is, Euclidean. For this to happen, there has to be a lot more mass in the universe than we can either see or compute from gravitational studies of galaxies. The flat-geometry models of most inflationary and related cosmological theories are built on the assumption that there is a critical density of mass for the universe and that if the actual average mass density in the universe is equal to the critical mass density, then the geometry is flat.

Theorists that espouse these views are looking for the missing mass. When neutrino mass was discovered, the finding raised hopes that the neutrino might hold the key to the missing mass. However, if neutrinos do have mass—and there are a *lot* of them in the universe—the additional mass is believed to still fall far short of the missing component. Either there are other huge sources of mass hiding in the universe, or the mass density of the universe is too small. If it is smaller than the critical mass, the universe is predicted to expand forever. Only if the mass density is greater than the critical mass density may the universe collapse back on itself due to gravity and produce a big crunch, possibly leading to a new universe from another big bang to follow.

The questions about the mass density of the universe, whether the universe is dominated by mass or by something else, and whether missing mass is present, all lead to an important concept: the overall geometry of the universe. Einstein assumed in his original field equation that the universe is dominated by mass. When the cosmological constant was introduced, however, the door was opened for another possibility. The new model could account for both the effects of mass and gravity and the effects of something else—an unseen force countering gravity, some mysterious energy of the void. Einstein's equations inexorably deal with the nature of space: its geometry.

CHAPTER 14

The Geometry of the Universe

"God ever geometrizes."
—Plato

We now come to an interesting question: What is the overall geometry of the universe? We know that locally, near a star or another massive object, space is curved. Space curves *around* the object spherically, as was demonstrated by the eclipse experiments. But what is the *overall* shape of the universe? Geometry is directly connected with mathematical equations. From studying Einstein's field equation, we should get an idea about the geometry of the universe.

The geometry of the universe helps determine its ultimate destiny. Mathematicians have identified three possible geometries for the universe as a whole. The first is the flat, Euclidean geometry. The curvature of space in the Euclidean universe is defined as zero. Curvature is a concept named after Gauss and denoted by the letter k. The assumption here is that the universe is a surface of *constant* curvature everywhere. For a flat universe, we say that the curvature is $k=0$. Surfaces of constant *nonzero* curvature fall into two categories. Either the curvature is positive, which we denote by $k=+1$, or the curvature is negative, which we write as $k=-1$. A surface with curvature $k=+1$ is "closed." In two dimensions of space, this would be the surface of a sphere.

When a surface has curvature k=-1, it is "open" and the geometry is hyperbolic, as in the model of Gauss, Bolyai, and Lobachevsky. Here, in two dimensions of space, the surface of constant negative curvature is the outer surface of a *pseudosphere*. The three constant-curvature models in two dimensions are shown below.[1]

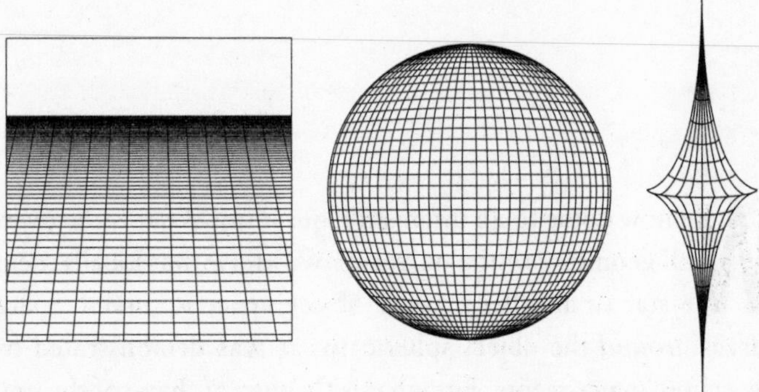

Let us consider the four dimensions of space-time, or equivalently, trace the evolution of a three-dimensional universe as it evolves through time, using the three possible constant-curvature geometries. We will now see why the words "flat," "closed," and "open" are used by cosmologists to define types of possible universes based on their shape. To do this, we will go to Einstein's field equation, which determines the geometry of the universe.

Einstein's equation, without the cosmological constant, is:

1. In the four dimensions of space-time, the special metric of relativity theory is used to define distances. With this metric, the hyperbolic space of negative curvature is often approximated by a saddle.

$$R_{\mu\nu} - 1/2 g_{\mu\nu} R = - 8 \pi G T_{\mu\nu}$$

But under the assumptions of a homogeneous, isotropic, constant-curvature universe, the tensor-valued equation (recall that the quantities $R_{\mu\nu}$, $g_{\mu\nu}$, $T_{\mu\nu}$ are tensors-they represent arrays of elements rather than simple numbers) above simplifies to a scalar (that is, non-tensor valued) differential equation written as:

$$(R'/R)^2 + k/R^2 = (8 \pi G/3)\rho$$

where ρ is the mass density of the universe. A differential equation is an equation that relates a derivative of a variable to several other quantities. Here, R is a scale factor that measures the size of the universe. Its derivative, R', measures the *rate of change in the size of the universe*. The equation, which is a simplification of Einstein's general field equation for the case of a "simple universe"—one that looks the same everywhere and in every direction—is thus a differential equation for the size, R, of the universe. The model here assumes a mass-dominated universe, that is, a universe where mass, rather than other forms of energy, constitutes the dominant force. This is what allowed us to replace Einstein's general energy-momentum tensor with a scalar quantity that measures mass.

When the possible values of k, being 0, +1, and -1, are inserted into the equation above, we get, respectively, that the mass density of the universe ρ is equal to, greater than, or less than $(R'/R)^2/(8 \pi G/3)$. This quantity is very interesting, and plays a crucial role in models of cosmology. First, the element R, the scale factor of the universe, measures the radius of curvature of the universe if the universe is closed and thus has such a curvature. The quantity (R'/R), the ratio of the derivative of the measure of size to the measure of size itself, is equal to Hubble's constant, denoted by H and measuring the expansion rate of the universe (that is, the rate of expansion measured in proportion to the size of the universe).

The entire quantity $(R'/R)^2 /(8 \pi G/3)$ is the critical density of the universe. We see that when the universe has exactly this density, that is, ρ is equal to the expression above, the curvature must be k=0, a flat universe. When ρ is greater than this expression, then k=+1. Here, the universe is heavier than the critical mass and therefore will eventually collapse into itself. When ρ is less than this critical density, the geometry is hyperbolic, as k=-1. Here, there is not enough mass in the universe and gravitational forces will not be strong enough to pull the universe together—it will continue expanding forever. In the case of a flat universe, the universe still will expand forever, but at a constantly decreasing rate.[2]

Cosmologists have a special name for the ratio of the two densities, the actual mass density of the universe at a given time, ρ, and the critical density given in the expression above. The ratio of the actual density to the critical density is called Ω (omega).

Omega holds the key to the geometry of the universe. Assuming no cosmological constant, the following is true. When omega is equal to 1, the density is equal to the critical density, and the universe is flat—it will expand forever, but will continually slow down its rate of expansion.

When Ω>1, the mass density of the universe is greater than the *critical* density at which the universe hangs in balance, and the universe slows its expansion. Here there is more mass than needed to just slow the expansion, and the universe will some

2. I am indebted to Jeff Weeks for the derivation of the geometry from the field equation.

day stop its expansion and start to contract to an inevitable "big crunch" that will swallow up everything. Then, perhaps there might be a rebirth with a new big bang, continuing the cycle of big bangs and big crunches as each new universe is built on the ashes of the previous one.

In the case $\Omega<1$, the mass density of the universe is smaller than the critical density. There isn't enough mass to stop the expansion and bring about a collapse, and the universe continues to expand forever. Its geometry is hyperbolic.

With a nonzero cosmological constant, the fate of the universe can be different in each one of the cases above, and it depends of the values of both parameters, Ω and λ.

The geometry of the universe is determined by how the three-dimensional universe evolves through time. A spherically-shaped universe that expands and then starts to collapse back on itself—a universe where $\Omega>1$—will trace a cycloid when mapped against time (here the universe at any point in time is represented by a circle; we suppress one spatial dimension in order to be able to draw this on paper). The picture of such a universe is shown below.

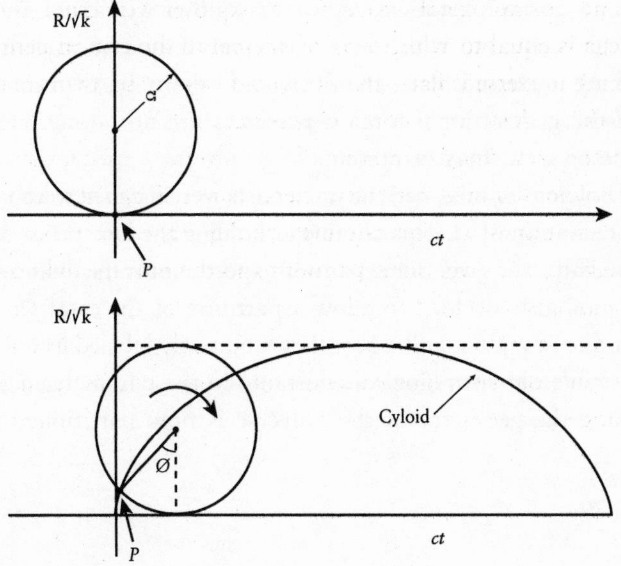

A flat universe, that is, one with $\Omega=1$, will expand at a decreasing rate.

A universe that has $\Omega<1$ will expand at an increasing rate as shown below.

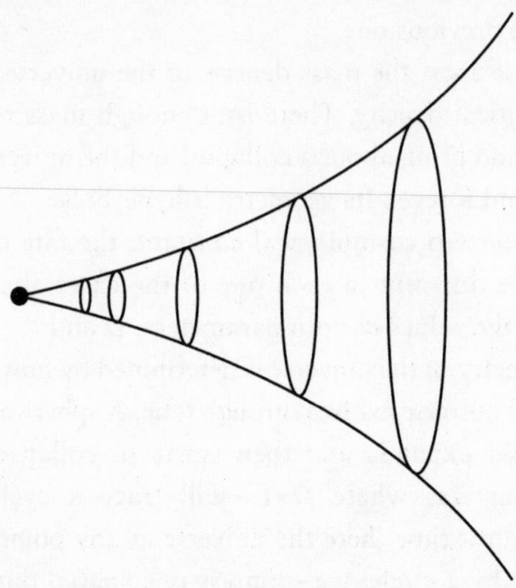

But what happens if there is something else in the universe that affects its expansion, geometry, and destiny? If there is some "funny energy" out in space, something that we cannot see or feel or detect, but which acts on the very fabric of spacetime, making it expand faster than it would otherwise, then matter and the gravitational force it produces are not alone. Here, something else may be present.

Allowing for this possibility, scientists were inclined to change the definition of Ω. This parameter holding the key to our universe had to be given some latitude to account for the unknown. Cosmologists decided to allow a partition of the total Ω: the part due to matter, and a part they tentatively labeled as due to Einstein's old cosmological constant. So the parameter determining the geometry of the universe is now partitioned as:

$\Omega=\Omega_M+\Omega_\Lambda$. While Ω determines the geometry of the universe, the fate of the universe is determined almost entirely by the "funny energy," Ω_Λ. The energy of Einstein's cosmological constant is potentially so overpowering that it could push the universe to expand forever almost regardless of what may be the value of the term Ω_M.

The possibility of determining both the geometry and the destiny of the universe was the motivation for the founding of the Supernova Cosmology Project in 1988 by Saul Perlmutter and his colleagues. These scientists wanted to use astronomical observations to try to estimate the values of the omega parameters. They would accomplish this task by studying the light curves of their "standard candles"—Type Ia supernovae. But the results this group of ingenious astronomers obtained went far beyond anyone's expectations. After collecting observations and performing calculations over a number of years, it became clear to the group that some unknown force of a magnitude never before observed or contemplated by science was at play in the universe. The value of Ω_M was smaller than anyone had expected. The exploding stars halfway across the cosmos were telling a strange and fantastic tale: there was not enough mass in the universe for any mass-based theory to hold, and there was an unseen force pushing everything apart faster and faster. Einstein's cosmological constant and the term corresponding to it, Ω_Λ, were immeasurably significant. But what were these findings telling us about the universe?

The Three Models of the Geometry and Destiny of the Universe

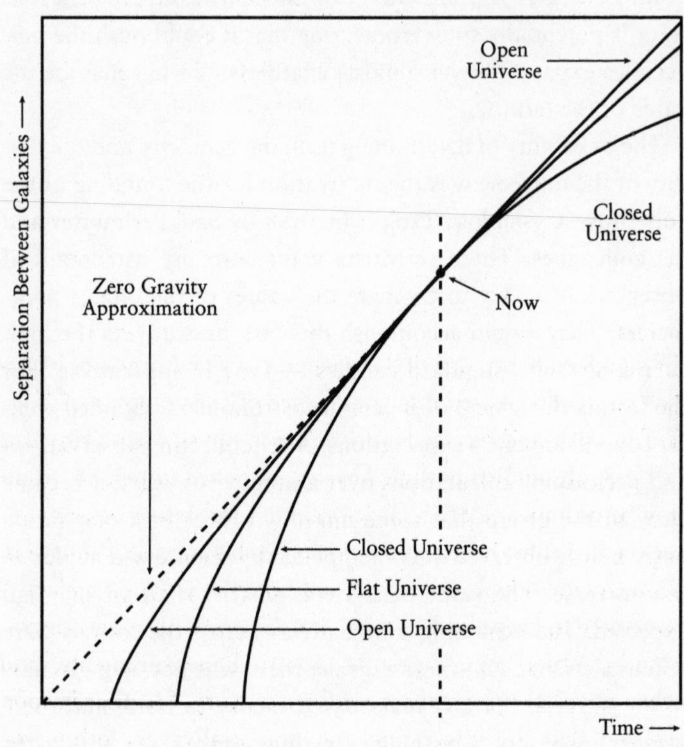

CHAPTER 15

Batavia, Illinois, May 4, 1998

"The universe is underweight."
—Neta Bahcall

Paul Steinhardt, one of the youngest of a generation of physicists who created the new science of cosmology, was born in 1952 and received a doctorate in physics from Harvard in 1978. He studied particle physics, but soon turned his attention to cosmology. Steinhardt became a professor at the University of Pennsylvania and studied Guth's model of inflation. Unlike many others, Steinhardt did not take inflation as fact—he wanted to keep an open mind and always let the data—astronomical observations, cosmic radiation measurements, and other physical information—tell their story. Steinhardt soon realized that the interesting and potentially very promising model of inflation had some theoretical problems. First and foremost among these was the unknown mechanism that stops inflation and brings in its wake the more tame expansion we believe takes place during our time.

Paul worked with a doctoral student, Andy Albrecht, and eventually solved the puzzle by revising Guth's inflationary model so that the expansion—and the field causing it—develop at a more inhibited rate, which still achieves the goals of inflation in explaining physical phenomena, but at the same time can be stopped by nature at some point. He then worked on a model called extended inflation, where another field interacts with gravity. In this theory, very early in the life of the universe, the gravitational constant was not the same as it is today. Thus the term *G* in Einstein's field equation was not a constant when the universe was still in its infancy. During this early epoch—called the Planck time and lasting the first 10^{-44} second from the big bang—quantum effects were present. At this very early stage in the life of the universe, quantum mechanics—the theory of the very small—actually had a say about what will happen to the entire universe, and Einstein's classical theory of relativity did not work here. A new field was needed, quantum cosmology, and Paul Steinhardt made contributions to our understanding of this important nascent theory.

In 1995, Steinhardt was pondering physical and astronomical results from various sources, which he felt were all pointing in one direction: the universe seemed to be accelerating its expansion. This was very counter-intuitive. Why should the universe be behaving this way when the theory of gravity—the only known long-range force in the universe—tells us that the expansion that started with the big bang should be slowing down as matter pulls on all other matter?

In September 1997, in response to the cosmological results, Paul Steinhardt decided that it might be a good idea to organize a meeting of practitioners in many fields related to cosmology: astronomers, astrophysicists, experimental physicists, particle physicists, applied mathematicians, and others, to discuss what was going on in the universe. What were the interpretations given by scientists in these diverse fields to new findings and their possible meanings? Steinhardt decided that the best place for such a meeting was the Fermi National Accelerator Laboratory (Fermilab, for short) in Batavia, Illinois. Here, many important experiments were being undertaken in an effort to understand the nature of matter under conditions simulating those that might have existed in the early universe. Together with Joshua Frieman of the Fermilab, Steinhardt made arrangements for a meeting to take place in May, 1998.

In January 1998, data on the first eight supernovae studied by Perlmutter and the Supernova Cosmology Project group were published in the journal *Nature*.[1] The data seemed to imply that faraway galaxies in space and time, such as the eight studied by the group, might be receding slower than nearby galaxies. The group had observations of sixty other galaxies with Type Ia supernovae still to be analyzed. Would this trend be confirmed by the data from the other galaxies? If so, these observations would provide more evidence that the universe was expanding faster during our time than it did in the past.

In the January 1998 meeting of the American Astronomical Society in Washington, D.C., Perlmutter's team presented their results suggesting that the universe might be expanding faster than expected. The competing Harvard-Smithsonian supernova group later reported limited results in agreement with Perlmutter's hypothesis. A Princeton University group headed by Neta Bahcall, and another Princeton group headed by Ruth Daly, reported results supporting this hypothesis as well. Their studies indicated that the total mass in the universe was not enough to ever stop the universal expansion.

Neta Bahcall was born and raised in Israel. She studied mathematics and physics at the Hebrew University, and earned a Masters in nuclear physics from the Weizmann Institute in 1965.

1. Perlmutter, S., et al., "Discovery of a Supernova Explosion at Half the Age of the Universe," *Nature*, Vol. 391, January 1, 1998, pp. 51–54.

Neta Bahcall
Photograph by Robert P. Matthews, Communications Department Princeton University

That year, she met her future husband, John Bahcall, who was a professor of physics at Caltech and was visiting the Weizmann Institute. The two were married the following year and moved to Caltech, where Neta worked on her Ph.D. in nuclear astrophysics under the supervision of William Fowler, who some years later received the Nobel Prize. Her research concentrated on the nuclear reactions that take place inside stars and make them shine. In 1970, Bahcall received her Ph.D. from Tel Aviv University. She became very interested in astronomy, doing joint research projects with astrophysicists at Caltech, studying quasars and other astronomical phenomena.

In 1972, Neta and John Bahcall were observing stars through the telescope of the just-completed Wise Observatory in the

Negev Desert in Israel. They were given sleeping quarters near the observatory, but had no babysitter. So they brought their two children, Safi, 3, and Dan, less than 1 year old, with them to the observatory and made them beds out of drawers they pulled out and lined with blankets. While their children were asleep, Neta and John Bahcall discovered the first binary eclipsing optical system that was also the first X-ray binary system detected by satellite. The compact object causing gas to emit the X-rays was a pulsar—the first binary pulsar ever discovered. This was very big news in astronomy. The discovery was the first from the new observatory, and the State of Israel selected Neta Bahcall as the year's Woman in Science.

The Bahcalls moved to Princeton, where Neta is a professor of astrophysics at Princeton University and John is a professor of natural sciences at the Institute for Advanced Study. In 1998, John Bahcall received the National Medal of Science from President Clinton. For six years Neta Bahcall was Head of the Science Selection Office at the Space Telescope Science Institute, where she made great contributions to science by helping select important astronomical projects for the Hubble. Over the years, her interests shifted to cosmology, and she became increasingly interested in how astrophysical discoveries can shed light on the structure of the universe, its beginning, age, and ultimate destiny. Neta Bahcall spent many years studying the large-scale structure of the universe in an attempt to answer the questions of cosmology. Her research was very fruitful, and based on her discoveries she was elected in 1997 to the U.S. National Academy of Sciences.

At the January 1998 meeting of the American Astronomical Society, Neta Bahcall presented results based on a number of studies she and her colleagues had completed, using several independent methods of "weighing" the universe. The researchers used clusters of galaxies to study the evolution of matter in the universe and its distribution. One of the methods employed Einstein's gravitational lensing effect. Here, the light from faraway galaxies was observed as it bent around galaxies that lie closer to us, and the amount of the bending of light provided information about the mass of the closer galaxies. Other methods studied hot gases within galaxies as well as velocities and redshifts and the ratio of mass to light in the universe. Bahcall studied the halos of galaxies, where much of the dark matter in the universe resides, as her research has indicated. Based on all of these studies, Neta Bahcall has concluded that the density of mass in the universe is only 20% of the density that would be needed to effect a slowing-down of the expansion and an eventual collapse. This number was obtained independently using several research methods. The probability that the conclusion is wrong was computed to be less than one in a million.[2]

Much was said in the press about the stunning new findings of Perlmutter, Bahcall, and their colleagues. The new revelations captured the world's imagination. Without saying it in so many words, it seemed that everyone—from scientists to the average citizen—had hoped for a "contained" universe, possibly a remnant of Einstein's static model. If the universe is indeed expanding and not stationary—as Hubble's research in the 1920s first told a surprised world—then people wanted it at least to oscillate between expansion and contraction. An alternately expanding and contracting universe could hold a possibility of an ultimate renewal, albeit in a dauntingly faraway time in the future. A universe expanding forever with no hope of ever contracting and returning to another big bang beginning was an

2. Statistical p-values were less than 10^{-6}.

unsettling scenario. So when the May meeting at the Fermilab took place, the press was there.

What concerned the scientists who gathered at the Fermilab was even more fundamental than an ever-expanding universe. It was the fate of physics. These scientists were facing an almost unavoidable conclusion: there was something very weird going on in the universe—something that no scientist could understand. Nature had a fifth force in its arsenal, one that had never been directly observed. This feeling, a guess shared by all who were present, physicists, particle theorists, astronomers, coalesced as technical explanations of the findings were presented. Scientists are trained to be skeptical; they want to see convincing evidence before they agree to toss out an old theory in favor of a new one. As everyone came together, 60 people in all, the dramatic presentations began.

William Press, an astronomer with the Harvard-Smithsonian group, played devil's advocate to his team and the Berkeley team, as the two supernova groups presented their findings. "What if the results are not right?" was the big question. Saul Perlmutter and Robert Kirshner, a member of the competing group, defended their teams' findings. There were many potential problems with the data. First, are the supernovae really "standard candles"?[3] How do we know that a Type Ia supernova that took place seven billion years ago has the same lightcurve as one that took place only half a billion years ago? Then there was a question about a correction that the teams applied

3. By the spring of 1999, a number of research papers in scientific publications have confirmed that the supernovae can well be taken as standard candles with good accuracy and that the resulting estimates of distance and speed can be given a high degree of confidence.

to the luminosity data so that the data could be compared. How did the correction work? Finally, there was the problem of the unexpected lack of dust in the galaxies studied. Why was no dust detected?

The two teams launched into lengthy technical explanations of the details, and they seemed to answer the questions to everyone's satisfaction. A straw vote was called, and the scientists voted almost overwhelmingly in favor of accepting the new information as scientifically convincing. Now there was no escape from the inevitable, haunting question: What was driving the accelerating expansion of the universe? The total mass of the universe should be pulling its weight, so to speak. An expansion from the big bang should somehow be slowed down by the existence of the mass distributed in galaxies throughout the universe. But clearly the evidence presented at Fermilab that day in May 1998 seemed to indicate that nothing of the sort was happening. The universe does not contain enough mass to slow down the expansion. And some mysterious force is actually *accelerating* the expansion. There was some *negative pressure* in the vacuum—something totally alien to all of science. Or was it?

"There is some funny energy in the universe," wrote Michael Turner of the Fermilab on his notepad. He added a drawing of stars and people scratching their heads, and a big Greek letter: Λ. His drawing found its way to the front page of the Science Section of the *New York Times* (May 5, 1998). "What was good enough for Einstein," Turner said, "ought to be good enough for us," referring to the cosmological constant.

But cosmologists involved with the new theories, such as inflation, wanted to go a step further. The cosmological constant can be used, in principle, to account for the mysterious force of nature pushing out on space, countering gravity and making the universe accelerate toward infinity. But according to inflation, there was once a similar force in the universe, and it made the universal expansion race at an exponential rate during the first tiny fraction of a second following the big bang. So the cosmological constant should be good for that era as well. But a problem arose here. The magnitude of the invisible force had to have been different during the primordial instant following the big bang from its magnitude during the present time. How can science incorporate a varying Λ?

A natural answer to this important question, which might solve a lot of mysteries of modern cosmology, would be to make the cosmological constant a cosmological *variable*—a function of time or other variables in Einstein's equation. But no one knows how to do this. Einstein has been dead for over four decades, and it seems that no one had the courage, insight, and knowledge to alter his equation the way Einstein himself had done when he inserted the cosmological constant in the first place.

Einstein's followers, the physicists who specialize in general relativity, spend their time solving Einstein's field equation. For this purpose, they use a battery of modern and old methods: some of them numerical techniques implemented on a computer, others theoretical derivations for solving complex differential equations. But these physicists do not attempt to change Einstein's equation to fit new findings and new theories.

Einstein's field equation is an icon. The equation was crafted by a great master. Every tensor, every constant, every minute element is there for a reason. The tensor equation is designed to preserve the laws of nature. These laws are invariant—they do not change when one looks at a physical process from a different angle, or a different coordinate system. In the limit, the tensor equation gives the simpler Newtonian laws that apply at non-relativistic conditions.

Einstein was able to include what he would later consider the rogue term in the equation by shrewdly manipulating the metric tensor, teasing it, bending space just a bit to accommodate the constant without losing the properties he had worked so many years to impart to his equation.

But making a mere constant into a new variable? Perhaps even the great master would not have been able to accomplish such a feat. So the cosmologists who thought it might be a great idea to explain the "funny energy" of the universe and also use the opportunity to buttress the inflationary theory attempted to do the next best thing: invent a new concept.

Paul Steinhardt pursued one such alternative model. He called it *quintessence,* after Aristotle's fifth element of nature. The name for the invisible force is an oblique reference to a fifth force of nature. The first four forces known to physics are gravity, electromagnetism, and the weak and the strong nuclear forces. Quintessence, which no one has yet observed, would be the fifth. Steinhardt is currently looking at ways to build quintessence into Einstein's field equation. Whatever theory might work in the end, Steinhardt summed up the present mystery. "There is negative pressure in the universe," he said to me. "One thing is clear now, and that is that Ω_M is less than 1. What does this mean? Curvature, quintessence, Λ?—We don't know, but whatever it is that's out there has consequences to fundamental physics."

CHAPTER 16

God's Equation

"I want to know God's thoughts."
—Albert Einstein

Einstein's cosmological constant had never really died, even if its inventor had washed his hands of the constant. In his monograph, Steven Weinberg traces the elusive constant's adventures.[1] Weinberg shows how adding the constant to Einstein's equation contributes a term equal to $\Lambda/8\pi G$ to the total effective energy of the vacuum. The question is whether the constant reflects all the energy in the vacuum, or whether there is something more pushing our universe outwards. And if the cosmological constant bears sole responsibility, how large must it be?

In the 1960s and 1970s, particle physicists were interested in the cosmological constant, since they had to estimate the energy levels of empty space in order to distinguish such energy from those of the particles they were studying in their accelerators. But however hard they tried, particle physicists could not make the amount of energy they expected to be present in empty space match anything that could be provided by the use of the cos-

1. Steven Weinberg, "The Cosmological Constant Problem," Morris Loeb Lectures in Physics, Harvard University, May 1988.

mological constant. Consequently, the particle physicists abandoned their attempts. Around that time, however, cosmologists rediscovered the disfavored constant and tried to recruit it for their own use. A pressing problem in cosmology in the late 1960s had no clear solution. This was the problem of the quasars.

Quasars (or quasi-stellar objects) release large amounts of radio energy, which can be detected by astronomers. An inexplicably large number of quasars were discovered at redshifts of about $z=1.95$. These quasars were all very distant from us in time and space, and were almost all created at about the same time (as their redshifts, indicators of their speeds of recession from us, clearly establish). But why? Cosmologists knew that the phenomenon could be explained if, somehow, the universe did not expand much during the time the quasars were formed, placing them at about the same distance from us. So what was desired was a universe that loitered for a while at a given size, R, corresponding to what it should be at the estimated age of the quasars. One way to make the expansion of the universe slow down, or even stop, at a given time is by applying the cosmological constant. Some cosmologists spent years attempting to make the cosmological constant give the right answer to this mystery.[2]

Next came more particle physicists. This time, they tried to answer questions about spontaneous symmetry breaking in the electroweak theory. Broken symmetries are the mechanisms by which particles of various types were believed to have been created in the early universe. Particle physics says that since an electron is different from a quark, a "symmetry" had to be broken

2. The phenomenon of the quasars has other explanations as well. I am indebted to Alan Guth for this observation.

to create the two different particles. The theorists faced a problem in which certain density calculations gave negative numbers. At some point it occurred to the theorists that applying the cosmological constant Λ to their equations would cancel a crucial term, answering the key question with a positive answer. A side effect, however, was the implication that in the distant past Λ had to have been very large. The deduction was puzzling until Guth came up with the inflationary universe theory. If the cosmological constant was indeed much larger in the very short instant right after the big bang, then it could have been the force driving the universe's exponential expansion during that epoch. Thus, adding the constant to the equation meant we had to choose the value of the constant very carefully.

Science needed fresh new theories, requiring new mathematics. It was not enough to tinker with Einstein's equation by trial and error. The remarkable equation had performed tremendously well over the years, always leading to new physical discoveries in amazing agreement with the equation's predictions. But when one tried to work with the equation including the cosmological constant—or to attempt to wed the relativistic equation to quantum theory—the results were poor. Humankind's basic understanding of the magic equation was simply not great enough.

In 1985, came a theory that seemed to hold great promise in leading to the solution of many problems in physics: the superstring theory. The theory extended the four dimensions of spacetime to eleven dimensions, in the sense that equations in superstring theory that attempt to model the universe use eleven dimensions. The theory's results were intriguing, but they have not yet solved the problem of Einstein's cosmological constant. In the late 1980s, mathematicians developed a variant of the superstring theory with the compacting of two space variables. But when they tried to extend the result to four-dimensional space-time, the technique failed and the whole structure collapsed.

In December, 1996, newspapers in London reported that the famous cosmologist Stephen Hawking was "going on a crash course in mathematics."[3] A professor of mathematics at Oxford was scheduled to talk about the topology of four-dimensional surfaces and their relations to general relativity and the quantum theory. Hawking and other cosmologists were very interested. The mathematician had apparently discovered a strange relationship between four-dimensional surfaces and "exotic" physical phenomena, unique to four-dimensional spaces. Sir Roger Penrose, himself a renowned topologist who had successfully applied the abstract mathematical discipline to physical problems described the finding: "What he did was to use ideas about the behavior of fundamental particles to establish a result in pure mathematics which was totally unexpected. Unique among the dimensions, four is the only one with this property." The mathematician's work in pure mathematics may have given a glimpse of the reason cosmologists and physicists have had such a hard time with the cosmological constant and other problems: Four-dimensional geometry is pathologically "badly behaved," uniquely so among dimensions. Apparently, we are cursed to live in a four-dimensional universe (three dimensions for space plus one for time), at least as far as physics is concerned.

Undaunted by the new findings, and seeing in them a new opportunity, cosmologists set to work, weaving the results about the topology of four-dimensional spaces into their own theory relating relativity and the quantum theory. Their aim was more

3. H. Aldersley-Williams, "May the Force be with Us?" *The Independent*, December 2, 1996, p. 20.

than to nail down the cosmological constant—it was to try to achieve the physicist's most lofty goal: a unified field theory, which would wed all the forces of nature. In trying to reach this aim, cosmologists were following in the footsteps of Einstein himself. For Einstein spent the remaining years of his life, after immigrating to the United States in 1932 and joining the Institute for Advanced Study, trying to unify the various areas of physics.

Albert Einstein pursued his scientific quest for knowledge with great passion. He was a sincere believer, and to him science was the process of discovering God's creation. Many of our greatest scientists today are driven by a similar quest. They are the front line of investigators looking to solve the puzzle of creation. These scientists ask deep philosophical questions about where the universe came from, where is it likely to go, and what is its shape.

In 1997, Stephen Hawking said that he was confident that within twenty years, we would understand the ground rules of the universe. Rival cosmologists quickly pointed out that twenty years earlier, Hawking made the same prediction. But Hawking acted as if he had something up his sleeve. Echoing Einstein, he said: "We are nearing God."[4] In March, 1998, Hawking revealed a bit of his hand. He had used the concept of an *instanton* to attempt to explain the big bang. Hawking and his co-workers said that this concept brought science closer to a "theory of everything," and a month later, in typical high drama, announcements of the new theory were made simultaneously by Hawking, while visiting California, and by his collaborator Neil Turok, back in London. But Hawking is not Einstein, and his invoking God, as Einstein used to do, may not

4. *The Observer*, November 23, 1997.

have been appropriate. To date, Hawking and his collaborators have not produced a single theory that can compare with Albert Einstein's theories.

What have these scientists accomplished so far? Hawking and Turok started with Alan Guth's inflationary model. Anticipating the discoveries announced in 1998 about the universe being "open" and thus infinite in the amount of space it will eventually fill as it expands forever, they asked the following question: Does inflation necessarily entail a "flat" universe, as most inflation supporters believe, or can it produce an open universe? Hawking and his collaborator James Hartle had tried a similar approach a few years earlier, applying inflation to a closed universe, using the technique of a path integral invented by the legendary American physicist Richard Feynman.

In 1995, Turok, who did not believe in flat or closed universes, was giving a talk in Cambridge about research results related to an open universe. His talk caught Stephen Hawking's attention, and the two began to work together. They tried to apply the approach taken by Hartle and Hawking to an open universe, but did not succeed for a long time. The problem was infinity. The appearance of an infinite component in the equation prevented the Feynman path integral from being valid. One day, Turok was writing mathematical expressions on Hawking's blackboard. Suddenly, Hawking stopped him by calling to him through the computer that acts as his voice (Hawking is paralyzed except for one finger, which operates his computer mouse, allowing him to talk using a machine voice). Turok had apparently made an error by neglecting a term in the equations, a term that was actually significant. The two worked on the corrected expressions and miraculously the infinite component disappeared. The space they now had in front of them described the evolution of the universe from the big bang through inflation and onward to an open universe—without a singularity at its inception.

Instead of a singularity, Turok and Hawking proposed the instanton—a particle of highly-compressed space and time, having the mass of a pea but a size a millionth of a trillionth of a trillionth that of a pea. The instanton is called so because it is a particle that exists for only an instant. Before the instanton, time does not exist, and neither does space. Unlike the big bang singularity, the instanton is smooth. And as the instanton explodes, cosmic inflation begins just as predicted by Alan Guth. Eventually the universe that sprang out of the instanton expands forever. Time, and observations from space in the coming years, will perhaps tell us which of these theories is correct, how our universe began, whether it is possible to unify the forces of nature, and whether the cosmological constant is part of the overall equation of physics.

Most scientists agree that the universe must have begun with a tremendous expansion from a very hot dense state, some kind of a big bang. This tremendous initial expansion started creation—the formation of matter and energy and then galaxies, stars, and planets, as well as the mysterious dark matter that can't be seen. But there, after the big bang, the divergence in views and philosophical bents begins. Saul Perlmutter, the man who through the world's most powerful telescopes and instruments came closer than any human being to actually seeing the universe expand, is a careful scientist. And his findings bring him to one particular theory.

Saul Perlmutter was born in Urbana-Champaign, Illinois, in 1959. Both of his parents were academics, employed by the University of Illinois. When he was still a young child, the family moved to Philadelphia, where Saul attended the Quaker Friends

Saul Perlmutter
Lawrence Berkeley National Laboratory

School. As a student, he was always good in mathematics and science, so he spent more time and effort studying the humanities—they offered a greater challenge. He also played the violin. Saul went to Harvard, and graduated in 1981 with a degree in philosophy and physics. Then he went to the University of California at Berkeley to pursue a Ph.D. in physics. Here, Saul Perlmutter moved among several high-powered research groups over a number of years. In 1982 he started doing research with a group studying fractional charge particles, but soon moved on to work with a team of graduate students supervised by Professor Richard Muller. Muller was a close associate of the late maverick physicist Louis Alvarez, who together with his son Walter Alvarez found evidence that the extinction of the dinosaurs 65 million years ago was due to the impact of a large asteroid.

Muller and his students were pursuing this idea further: they were using astronomical observations of dim red stars to look for the star called Nemesis. Nemesis is a putative companion to the Sun, orbiting it once every 52 million years. When Nemesis is closest to the Sun, every 26 million years, mass extinctions occur because the gravitational pull of Nemesis hurtles asteroids in our direction. Finding Nemesis, if it indeed exists, would answer not only the question of how the dinosaurs died, but also why the catastrophe—and others like it—happened. The group measured the distances (using the parallax method) to about 300 stars out of a total group of 2,000 suspects obtained from a star catalog, but all were found to be too far away. Then the search was suspended for various reasons. Saul disappeared for a month, spending his time in the basement of one of the physics buildings at Berkeley. He reemerged with a new invention—a robotic telescope.

Looking for uses for his new, computer-controlled telescope, Saul came up with his ingenious technique of detecting supernovae by a systematic search of distant galaxies. By 1985 he was able to obtain electronic images of twenty of these rare explosions in faraway galaxies.

In 1986, Saul received his Ph.D. in physics from U.C. Berkeley, and the following year he and his colleague Carl Pennypacker decided that Saul's technique of finding supernovae using the robotic telescope could be used to measure the rate of *deceleration* of the universe. Their decision, as late as 1987, reflected the prevailing view in physics at that time: that our universe, which began with a big bang, had to be decelerating due to the mutual gravitational pull of all its mass. The team began taking measurements using observations from a four-meter telescope in Australia. They found supernova explosions at z=0.3, but the pace was slow. Weather didn't allow for frequent observations, and there were other difficulties as well. The team, which was now growing in size, moved its center of observations to La Palma, in the Canary Islands, where it had the use of a 2.5 meter telescope. The scientists were now up to a z-value of 0.45, and they developed a batch technique, which allowed them to obtain and process larger sets of galaxy observations, finding more and more supernovae. They were able to send their results directly to Berkeley via the Internet.

By now Saul had taken up a research position at the Lawrence Berkeley National Laboratory, in the hills above the Berkeley campus, and he made the lab his center of operations. With their increasing success, team members obtained the use of larger telescopes. Eventually they got to use the twin ten-meter Keck telescopes in Hawaii, the largest on Earth, as well as the Hubble Space Telescope. Results of distant explosions were streaming in with amazing regularity. But the picture all these observations were painting was exactly the opposite of what science had been expecting. The universe was not decelerating its expansion—it was picking up pace. Based on observations at various z-levels, it now seemed that from the time right after the big bang to about seven billion years ago the universe was indeed slowing down its expansion. But the density of matter in the universe was simply not enough to slow the expansion to a halt. As the universe continued to grow, its mass diluted, allowing Einstein's "funny energy" to take over. Seven billion years ago, the expansion rate thus started to pick up speed, and the universe is now expanding faster all the time.

As an experimentalist, Perlmutter is always open to all possible explanations of his data. Every alternative to a given hypothesis must be explored to its fullest extent, he maintains. But as new observations were collected, by the spring of 1999 it was clear that all of these measurements were indicating that the universe is expanding ever faster. Based on his observations, Perlmutter now believes that the universe is probably flat—its geometry is Euclidean—and that it will expand forever. By this time, Perlmutter’s group had collected data on galaxies outside the original range. The group studied the redshifts and distances of galaxies that are so far away ($z=1.2$) that by the time their light left on its way to us the universal expansion was still slowing down. This is in contrast with the bulk of the team’s data, consisting of galaxies in the range indicating the present accelerating-expansion universe (galaxies with $z=0.7$). The team collected data outside the original range—as all good scientists should do—in order to test the limits of their proposed theory. And so far the theory has withstood all the tests.

Based on the team’s findings, Perlmutter also believes that the cosmological constant is important and that rather than being Einstein’s “greatest blunder,” the constant is an integral part of the equation that defines the universe, its past and its ultimate fate. Perlmutter, like most other leading astronomers, also believes in the inflationary universe theory. Thus, based on the latest astronomical observations and the most widely accepted theoretical derivations, our universe began in a vast expansion of space; this expansion slowed down for several billion years, and then accelerated again and continues to do so. If these conclusions are correct, the universe will expand forever.

“We really don’t know what happened there—the big bang was a totally amazing occurrence. I don’t believe any of these theories about fields we haven't found or baby universes we have no evidence for, or a larger universe in which ours is embedded. There is no objective reason to believe in any of these hypotheses," Sir Roger Penrose said to me. In 1965, Penrose published his theorem, which, using the powerful methods of topology, provided evidence for the big bang singularity of space-time. "I believe the universe has a hyperbolic geometry, but I don't know about the cosmological constant—I don't believe in it. As for the inflationary universe theory—I am a skeptic. What these people do is come up with a theory, and when the evidence doesn't support it, they change their theory, then change it again and again."

Alan Guth counters these arguments with the following: "The details of exactly how inflation worked are still unclear, but I think that the basic idea of inflation is almost certainly right. Inflation offers the only convincing explanation I have heard for how the universe became so big, so uniform, and so flat."

The debate among scientists will undoubtedly continue, as they try to unravel the mystery of the universe. But there is one thing on which all of these scientists agree, and that is the sheer power and the ever-continuing usefulness of Einstein's general theory of relativity. In the final analysis, knowing "God's thoughts" more completely would require incorporating into the theory of relativity also quantum considerations. But whatever the final equation may be, Einstein's field equation will form a major part of it. In developing his amazing equation, Einstein realized his life's dream—he heard at least some of God's thoughts. This is Einstein's field equation with the cosmological constant, which is our best estimate of God's Equation:

$$R_{\mu\nu} - 1/2 g_{\mu\nu} R - \lambda g_{\mu\nu} = -8 \pi G T_{\mu\nu}$$

where $R_{\mu\nu}$ is the Ricci tensor, R is its trace, λ is the cosmological constant, $g_{\mu\nu}$ is the measure of distance—the metric tensor of the geometry of space, G is Newton's gravitational constant, $T_{\mu\nu}$ the tensor capturing the properties of energy, momentum, and matter, and 1/2, 8 and π are numbers.

In his book *Out of My Later Years* (New York: Philosophical Library, 1950, p.48), Einstein gave hints about how he saw the future, and why he had been unable to develop a unified theory of everything. He wrote:

> The general theory of relativity is as yet incomplete insofar as it has been able to apply the general principle of relativity satisfactorily only to gravitational fields, but not to the total field. We do not yet know with certainty by what mathematical mechanism the total field in space is to be described and what the general invariant laws are to which this total field is subject. One thing, however, seems certain: namely, that the general principle of relativity will prove a necessary and effective tool for the solution of the problem of the total field.

Einstein understood that his efforts had been limited by the availability of mathematical methods. In developing special relativity, Einstein used the mathematics of Lorentz and Minkowski. For general relativity, he successively used the mathematics of Ricci and Levi-Civita and that of Riemann. But here, Einstein had to stop. He had come a long way toward discovering God's Equation, but to go further, he would have needed new mathematics. Such mathematics will likely be found in the direction suggested by S.S. Chern in his Princeton address and will include abstractions of geometry and topology to higher levels. Mathematicians will develop the tools, physicists will apply them, astronomers will verify the theories and provide data, and cosmologists will generate the big picture of our universe.

Once each discipline is supported by developments in the others, we may begin to understand the ultimate laws of nature and to formulate our human estimate of God's Equation. When the final equation is constructed, we should be able to use it to solve the wonderful riddle of creation. And perhaps that's why God sent us here in the first place.

References

Bohm, David, *The Special Theory of Relativity*, Reading, MA: Addison-Wesley, 1979.

Bonola, Roberto, *Non-Euclidean Geometry*, New York: Dover, 1914. Includes the original papers by J. Bolyai and N. Lobachevsky.

Borisenko, A. I., and Tarapov, I. E., *Vector and Tensor Analysis*, New York: Dover, 1968.

Born, Max, *The Born-Einstein Letters*, New York: Walker, 1971.

Born, Max, *Einstein's Theory of Relativity*, New York: Dover, 1965.

Brian, Denis, *Albert Einstein*, New York: Wiley, 1997.

Calaprice, Alice, ed., *The Quotable Einstein*, Princeton, NJ: Princeton University Press, 1996.

Calder, N., *Einstein's Universe*, New York: Penguin, 1988.

Chandrasekhar, S., *Eddington: The Most Distinguished Astrophysicist of His Time,* New York: Cambridge University Press, 1957.

Clark, Ronald, *Einstein: The Life and Times*, New York: Avon, 1984.

Davies, P., *About Time*, New York: Simon & Schuster, 1995.

Demetz, Peter, *Prague in Black and Gold*, New York: Hill and Wang, 1997.

Do Carmo, M., *Differential Geometry of Curves and Surfaces*, Englewood Cliffs, NJ: Prentice-Hall, 1976.

Eddington, Sir Arthur S., *Space, Time and Gravitation: An Outline of the General Relativity Theory*, New York: Harper & Row, 1959.

Eddington, Sir Arthur S., *The Mathematical Theory of Relativity*, New York: Cambridge University Press, 1923.

Einstein, Albert, *Autobiographical Notes*, La Salle, IL: Open Court, 1992.

Einstein, Albert, *Out of My Later Years*, New York: Philosophical Library, 1950.

Einstein, Albert, *The Origins of the General Theory of Relativity*, Glasgow: Jackson, Wylie, 1933.

Einstein, Albert, *The Principle of Relativity*, New York: Dover, 1952. Includes papers by H. Lorentz, H. Weil, H. Minkowski, and notes by A. Sommerfeld.

Einstein, Albert, *Relativity: The Special and the General Theory*, New York: Crown, 1961.

Ferris, Timothy, *Coming of Age in the Milky Way*, New York: Anchor, 1988.

Fölsing, Albrecht, *Albert Einstein*, New York: Penguin, 1997.

Frank, Philipp, *Einstein: His Life and Times*, New York: Knopf, 1957.

French, A. P., ed., *Einstein: A Centenary Volume*, Cambridge, MA: Harvard University Press, 1979.

Gilbert, Martin, *A History of the Twentieth Century*, Volume I: 1900–1933, New York: Morrow, 1997.

Goldsmith, Donald, *Einstein's Greatest Blunder?*, Cambridge, MA: Harvard University Press, 1995.

Golub, L., and Pasachoff, Jay M., *The Solar Corona*, New York: Cambridge University Press, 1998.

Gruning, Michael, *A House for Albert Einstein*, Berlin: Verlag der Nation, 1990.

Guggenheimer, H. W., *Differential Geometry*, New York: Dover, 1977.

Guth, Alan H., *The Inflationary Universe*, Reading, MA: Addison-Wesley, 1997.

Halliday, David, and Resnick, Robert, *Fundamentals of Physics*, Vols. I and II, 3d ed., New York: Wiley, 1988.

Hawking, Stephen, *A Brief History of Time*, New York: Bantam, 1988.

Hentschel, Klaus, *The Einstein Tower: An Intertexture of Dynamic Construction, Relativity Theory and Astronomy*, Stanford, CA: Stanford University Press, 1997.

Holton, Gerald, *Einstein, History, and Other Passions*, Reading, MA: Addison-Wesley, 1996.

Hoskin, Michael, ed., *The Cambridge Illustrated History of Astronomy*, Cambridge, UK: Cambridge University Press, 1997.

Huggett, S. A., et al., *The Geometric Universe: Science, Geometry, and the Work of Roger Penrose*, New York: Oxford University Press, 1998.

Klein, Martin J., et al., *The Collected Papers of Albert Einstein*, Volume V, Princeton, NJ: Princeton University Press, 1993.

Kragh, Helge, *Cosmology Comes of Age*, Princeton, NJ: Princeton University Press, 1996.

Landau, L., and Lifshitz, E., *The Classical Theory of Fields*, Oxford: Pergamon Press, 1962.

Levy, S., ed., *Flavors of Geometry*, New York: Cambridge University Press, 1997.

Lightman, Alan, and Brawer, Roberta, *Origins: The Lives and Worlds of Modern Cosmologists,* Cambridge, MA: Harvard University Press, 1990.

Meserve, B., *Fundamental Concepts of Geometry*, New York: Dover, 1983.

Misner, C. W., Thorne, K. S., and Wheeler, J. A., *Gravitation*, San Francisco: Freeman, 1973.

Pais, Abraham, *Einstein Lived Here*, New York: Oxford University Press, 1994.

Pais, Abraham, *'Subtle Is the Lord . . . ': The Science and the Life of Albert Einstein*, New York: Oxford University Press, 1982.

Pasachoff, Jay M., *Astronomy: From the Earth to the Universe*, 5th Ed., San Diego: Saunders, 1998.

Penrose, Roger, *The Emperor's New Mind,* New York: Oxford University Press, 1989.

Rees, Martin, *Before the Beginning: Our Universe and Others*, New York: Helix Books, 1997.

Reichenbach, H., *The Philosophy of Space and Time*, New York: Dover, 1958.

Rucker, R. B., *Geometry, Relativity, and the Fourth Dimension*, New York: Dover, 1977.

Sayen, J., *Einstein in America*, New York: Crown, 1985.

Schatzman, E., *Our Expanding Universe*, New York: McGraw-Hill, 1992.

Spielberg, N., and Anderson, B., *Seven Ideas that Shook the Universe*, New York: Wiley, 1987.

Stachel, John, et al., *The Collected Papers of Albert Einstein*, Vols. I and II, Princeton, NJ: Princeton University Press, 1987, 1989.

Stoker, J. J., *Differential Geometry*, New York: Wiley, 1969.

Thorne, Kip S., *Black Holes and Time Warps: Einstein's Outrageous Legacy*, New York: Norton, 1994.

Weinberg, Steven, *Gravitation and Cosmology: Principles and Applications of the General Theory of Relativity*, New York: Wiley, 1972.

White, Michael, and Gribbin, John *Einstein: A Life in Science*, New York: Penguin, 1994.

Wolfe, H. E., *Non-Euclidean Geometry*, New York: Holt, Rinehart and Winston, 1945.

Index

Abraham, Max, 111–12
Adelhard of Bath, 48
Albrecht, Andy, 199
Alpher, Ralph, 176
Alvarez, Louis, 217
Alvarez, Walter, 217
Andromeda, 154, 163
Arab science, 47–48
Aristotle, 206
asteroids, 215
astral geometry, 58

Babylonians, 43–44
Bahcall, John, 200–201
Bahcall, Neta, 11, 199–202
Beltarmi, Eugenio, 101
Berlin, 66–68, 80–81, 105–8, 119–20, 125
Berlin Observatory, 36–37, 68
Besso, Michele Angelo, 17, 24, 28, 83
Bianchi, Luigi, 110, 114
Bianchi identities, 113–14
big bang, 168, 171–72, 217–18
  evidence for, 176–78
  instanton concept, 211, 213
  problems with standard theory, 173–75
  spacetime singularity, 168, 169–71, 213, 218
  See also expansion of universe; inflationary universe theory
black holes, 4, 119, 168–69
Bolyai, Johann, 56–59, 101
Bolyai, Wolfgang, 56, 58
Brazil, eclipse expedition in, 126–29, 133–34, 135, 137, 142–44
broken symmetries, 208–9

Cepheid variables, 161–62, 163
Chandrasekhar, S., 122, 146

chemical elements in universe, 177–78
Chern, S. S., 99–100, 219
Clark, Ronald, 82
Corry, L., 114
Cosmic Background Explorer satellite (COBE), 177
cosmic background radiation, 175, 176–77
cosmological constant
  challenges to, 158–60
  Einstein's abandonment of, 160, 164–65, 168
  Einstein's introduction of, 155–58, 167–68, 206
  as measurement of force expanding universe, 180, 204–6
  relationship to curvature of universe, 156–57
  relationship to expansion of universe, 193, 194–95, 208, 217
  relationship to inflationary universe theory, 175, 205, 209
  usefulness, 167–68, 204–5, 207, 217
  as variable, 205, 209
cosmology
  Einstein on, 149, 151–59
  flatness problem, 173
  horizon problem, 173–75
  quantum, 198
  search for unified field theory, 210–11
  See also universe
Cottingham, 134, 138
Cowan, Clyde, 182
Crab Nebula (M1), 2–3
Crimean eclipse expedition, 71, 79, 81–84, 85, 87
critical density of universe, 173, 187, 191–95
Crommelin, A. C. D., 126, 127, 143
curvature
  of finite universe, 156–57
  of flat Euclidean space, 94
  of hyperbolic space, 94, 101
  of spacetime, 59, 65, 94, 96, 118, 143–44
  of spherical space, 94
  tensor, 113–14
  See also light deflection

Daly, Ruth, 11, 199
dark matter, 186, 202
Demetz, Peter, 31
de Sitter, Willem, 121, 122–23, 140–41, 154, 158, 159–60, 165
Dicke, Robert H., 176
differential equations, 191
differential geometry, 97, 99–100
dinosaurs, 215
distances, in non-Euclidean

geometries, 95–97, 100, 117

Dyson, Frank, 124, 125, 138, 142–43

earth, believed to be flat, 43–44, 50–51

eclipses
- attempts to observe bending of starlight, 76–77, 78–79, 124–25, 147–48
- Brazilian expedition, 126–29, 133–34, 135, 137, 142–44
- Crimean expedition, 71, 79, 81–84, 85, 87
- Principe expedition, 124, 126–27, 129–34, 135–37, 138, 142–44

Eddington, Arthur, 90, 121–24, 125, 141
- on curvature of space and general relativity, 132–33, 143–44, 145, 146, 148
- poetry, 146–47
- Principe expedition, 126–27, 129–34, 135–37, 138
- research on expanding universe, 165

Ehrenfest, Paul, 68, 87, 111, 140–41, 158, 159

Einstein, Albert
- awareness of anti-Semitism, 30, 67, 107–8
- in Berlin, 66–68, 80–81, 82, 105–8, 119–20, 125
- as celebrity, 142, 147
- correspondence with Freundlich, 68, 73–79, 88, 89–90, 112, 147–48
- family, 14, 15–16, 19, 66, 80
- at German University (Prague), 29–31, 33–34, 38–39, 40
- Jewishness, 30–31
- knowledge of astronomy, 76–77
- knowledge of British confirmation of light deflection, 139–40, 141–42
- life, 14–19, 39
- marriages, 17, 19, 80, 107
- mathematical knowledge, 17–18, 61–63
- personality, 15, 40, 76, 108–9, 112, 115
- reputation, 119
- research and papers published, 21, 29, 38–39, 40, 119–20
- search for unified field theory, 100, 160, 211, 219
- work at Swiss patent office, 19, 21, 28, 59, 62
- in Zurich, 62, 65–66
- Zürich Notebook, 37, 116–17

Einstein, Elsa, 80, 107
Einstein, Mileva, 17, 66, 105
electromagnetism, 160
equations, 64–65
  differential, 191
  See also field equation of gravitation
Escher, M. C., 53, 54, 55
ETH. *See* Federal Institute of Technology
ether, 20–21, 22–24, 145
Euclidean geometry, 44–45
  fifth postulate, 45–51, 54–58
  of universe, 173, 186–87, 189, 191–92, 194, 212, 217
expansion of universe
  beginning of, 168, 171
  closed theory (collapse predicted), 6–7
  cosmological constant dropped because of, 160, 164–65, 168
  critical density and, 173, 187, 191–95, 216–17
  description, 178–80
  de Sitter on, 159–60
  distribution of matter, 186
  Einstein's initial doubts about, 153–54
  evidence for acceleration of, 7–10, 11, 199, 204–5, 216–17
  evidence for, 160–61
  "funny energy" as force in, 10–11, 180, 204–6
  Hubble's Constant, 164, 191–92
  Hubble's Law, 5, 164, 178
  models of, 165, 212
  steady state theory, 6
  theories, 6–7, 10
  varying rates of, 208, 209
  See also big bang; inflationary universe theory
extended inflation model, 198

Fan, Xiaohui, 11
Federal Institute of Technology (ETH; Zurich), 16, 17–18, 38, 40, 61–62, 63, 66
Fermi, Enrico, 182
Fermi National Accelerator Laboratory (Fermilab), 11, 198–99, 203–4
Feynman, Richard, 212
field equation of gravitation
  application to universe, 151–52, 156–57
  components, 114–15, 117, 151–52
  development of, 68, 75, 113–15
  early approximation of, 116–17
  geometry of universe and, 190–99
  as icon, 205–6

metric tensor, 65, 96, 151–52, 155
Nordström's work on, 75
power of, 13–14, 218–19
publication of, 117
solutions, 117–19, 205
See also cosmological constant
fields, 20–21
See also gravitational field
Fokker, A., 109
Fölsing, Albrecht, 24
Foucault's pendulum, 150
four-dimensional spacetime, 32–33, 190
fractals, 185
Frank, Philipp, 38, 40
Franz Ferdinand, Archduke, 84–85
Freundlich, Erwin Finlay, 36–37
astronomical research, 77–78, 111
career, 89–90
correspondence with Einstein, 68, 73–79, 88, 89–90, 112, 147–48
Crimean eclipse expedition, 71, 79, 81–84, 85, 87
relationship with Einstein, 71–79, 82, 87–88, 89–90, 147–48
research on light deflection, 74, 78–79, 147–48
Friedmann, Alexander, 165, 168
Frieman, Joshua, 199
functions, derivatives, 170

galaxies
dark matter in, 202
discovery of, 154, 163–64
distribution in universe, 184–86
formation of, 171–72, 177
measuring distances to, 4–5, 162, 163
most distant, 173–74
redshifts, 5
speed of recession, 4–5, 7–10, 160–61, 164, 178, 179
structures, 184–86
Galileo Galilei, 19, 21
Gamow, George, 176
Gauss, Karl Friedrich, 54–56, 58, 92, 93, 94, 102
Geller, Margaret, 185
general theory of relativity
application to entire universe, 152–54
development, 28, 31–32, 33–38, 39, 64, 107, 115–17
difficulty of understanding, 145–46
dissemination of Einstein's paper, 121, 122

- general theory of relativity (*continued*)
  - Eddington on, 132–33, 145
  - Einstein's desire for experimental confirmation, 68, 73–79, 82–83, 119
  - later work based on, 99–100, 112–13, 114–15, 141
  - mathematics of, 33–34, 62–65, 68, 96, 109–11, 112–14
  - meetings on, 140, 142–47
  - perihelion of Mercury and, 83–84, 109, 111, 145
  - power of, 218–19
  - reactions to, 82, 111–12, 144–45
  - reference frames, 32
  - *See also* field equation of gravitation; light deflection
- geometry
  - in ancient times, 43–44
  - differential, 97, 99–100
  - four-dimensional, 210
  - *See also* Euclidean geometry; non-Euclidean geometry
- geometry of universe
  - application of Einstein's field equation to, 190–99
  - curvature, 156–57
  - flat (Euclidean), 173, 186–87, 189, 191–92, 194, 212, 217
  - hyperbolic (non-Euclidean), 190, 192, 193, 218
  - non-Euclidean, 100–102, 144
  - possibilities, 189–95
  - relationship to cosmological constant, 156–57, 193, 194–95
  - spherical (non-Euclidean), 189, 192–93
- German University (Prague), 29–31, 33–34, 38–39, 40
- Germany
  - World War I, 85–87
  - *See also* Berlin
- gravitation
  - attempts to find unified theory, 100, 160, 212–13, 219
  - effects on galaxies, 186
  - Einstein's study of, 28, 64, 112, 117–18
  - expansion of universe driven by, 175
  - light bent by. *See* light deflection
  - Newton's theory, 27–28
  - redshift caused by, 32, 39
  - in special theory of relativity, 27–28
  - *See also* field equation of gravitation
- gravitational field
  - limiting value of, 151

space curved due to, 11, 59, 96
gravitational lensing, 37–38, 202
Grossmann, Marcel, 61–62, 63
collaboration with Einstein, 64, 65, 69, 75
criticism of general theory of relativity, 68–69
relationship with Einstein, 17, 19, 62, 68–69
group theory, 98–99
Guerra, Erick, 11
Guth, Alan, 159, 172–73, 175–76, 197, 209, 212, 213, 218

Haber, Fritz, 72
Hartle, James, 212
Hawking, Stephen, 210, 211–13
Herman, Robert, 176
Herz, Heinrich, 21
Higgs fields, 175–76
Hilbert, David, 112–13, 114–15
Hitler, Adolf, 81
horizon problem, 173–75
Hoyle, Fred, 168
Hu, Esther, 173–74, 179
Hubble, Edwin, 5, 161, 162–63
Hubble's Constant, 164, 191–92
Hubble's Law, 5, 164, 178
Hubble Space Telescope, 201, 216
Huchra, John, 185
Humason, Milton, 163
Hyades, 134
hyperbolic geometry, 52–54, 57–59, 100–101
curvature of space, 94, 101
of universe, 190, 192, 193, 218

inflationary universe theory, 159, 172–73, 175–76, 205, 213
cosmological constant and, 175, 205, 209
evidence for, 177, 217
extended model, 198
flat geometry of universe assumed, 186–87, 212
rate of expansion, 197–98, 211
skeptics, 218
See also big bang
instanton concept, 211, 213

Kamioka Neutrino Observatory, 183–84
Keck telescopes, 173, 216
Kirshner, Robert, 203
Klein, Felix, 98–99
Klein bottles, 98
Krupp, Gustav, 81

Lapparent, Valerie de, 185
Lawrence Berkeley National Laboratory, 1–2, 216
Leavitt, Henrietta, 161–62, 163
Lemaître, Georges, 165, 168
Levi-Civita, Tullio, 34, 63, 110
light
  gravitational lensing, 37–38, 202
  particle theory, 36, 89
  speed of, 21
light deflection by massive objects, 34–38, 39–40
  attempts to confirm, 36–37
  confirmation of, 133–36, 142–44, 147–48
  Einstein's desire for experimental confirmation, 36–37, 68, 73–79, 82–83, 119
  Einstein's estimates of, 88–89, 111
  See also eclipses
Lobachevsky, Nikolai Ivanovich, 58–59, 101
Lodge, Oliver, 144–45
Lorentz, H. A., 21, 140–42, 158

M1 Nebula (Crab), 2–3
Mach, Ernst, 19–20, 34, 149, 158–59
Mach's law of inertia, 149–50
Magellanic Clouds, 161, 163, 184
magnetic field, 20
magnetic monopoles, 175–76
Mandelbrot, Benoit, 185
Mandl, Rudi W., 37–38
massive objects
  collapse into black holes, 168–69
  See also light deflection by massive objects; stars
mass of universe
  average density, 157–58
  critical density, 173, 191–95
  less than critical density, 11, 183, 187, 199, 216–17
  measuring, 201–2
  missing, 184–87
  ratio to critical density, 192–95
  relationship to curvature, 156–57, 191–92
mathematics
  Einstein's view of, 17–18
  of general theory of relativity, 33–34, 62–65, 68, 96, 109–11, 112–14
  of special theory of relativity, 18, 32–33, 63
matter
  broken symmetries, 208–9
  creation in big bang, 171
  critical density of universe, 173, 187, 191–95
  distribution in universe, 184–87

neutrinos, 182–84, 187
particles, 181–82, 208
properties, 181, 186
Maxwell, James Clerk, 19, 20–21, 22–23
Mayr, L., 148
Mercury, perihelion of, 83–84, 109, 111, 145
metrics. *See* distances
metric tensor, 65, 96, 114, 117, 151–52
Michelson, Albert A., 22–24
Mie, Gustav, 112
Milky Way, 154
Minkowski, Hermann, 18, 32–33, 61, 63
Möbius, A. F., 98
Möbius strips, 98
Morley, Edward W., 23–24
Muller, Richard, 214–15

Nasiraddin (Nasir Eddin Al-Tusi), 47, 49, 50
nebulae
M1 (Crab), 2–3
spiral, 160–61
Nemesis, 215
Nernst, Hermann, 66–67, 108
neutrino astronomy, 184
neutrinos, 182–84, 187
neutron stars, 3–4
Newton, Isaac
gravitation theory, 27–28, 35, 125, 150–51
Mach's criticism of, 20
scattering theory, 36, 132
non-Euclidean geometries, 50, 100–102, 144
curvature, 94, 101
distances, 95–97
Einstein's use of, 15, 33, 40–41, 54, 59, 63–64
hyperbolic, 52–54, 57–59, 100–101, 190, 192, 193, 218
Riemann's work on, 94–97, 110
spherical, 51–52, 100, 189, 192–93
Nordström, Gunnar, 75, 112
Nöther, Emmy, 113, 114
nuclear fusion, 182

Omar Khayyam, 47
omega, 192–95

particle physics, 207–9
Pauli, Wolfgang, 182
Peebles, James E., 176
Pennypacker, Carl, 215
Penrose, Roger, 99, 168–69, 171, 210, 218
Penzias, Arno, 177
Perlmutter, Saul, 1–2, 4, 5–6, 7, 8–10, 162, 195, 199, 203, 215–19
Pick, Georg, 33, 34, 63

Planck, Max, 66–67, 79, 89, 90, 107, 108, 120
Playfair's axiom, 46, 57
Poincaré, Henri, 21
Poisson, Simeon-Denis, 150, 151
Pollak, Leo W., 36, 73
Prague, 29–31, 33–34, 38–40
Press, William, 203
Principe, expedition to, 124, 126–27, 129–34, 135–37, 138, 142–44
Proclus, 46–47
Prussian academy, 80–81, 82, 87, 105, 106, 110, 118
pseudospheres, 101, 190
pulsars, 3
Pythagoras, 44

quantum cosmology, 198
quantum theory, 39, 66, 106, 120, 181–82, 184
quasars, 208
quintessence, 206

radiation, cosmic background, 175, 176–77
redshifts, 5, 9
  caused by gravity, 32, 39
  as evidence for expansion of universe, 160–61
  of spiral nebulae, 160–61
reference frames, 21, 32, 59
Reines, Fred, 182, 183
relativity. *See* general theory of relativity; special theory of relativity
Renn, Jürgen, 37, 114, 116
Ricci, Gregorio, 34, 63, 110
Riemann, Bernhard, 69, 91–97, 99, 100, 102–3, 110
Russia, Crimean eclipse expedition, 71, 79, 81–84, 85, 87

Saccheri, Girolamo, 48–50, 51
Sauer, Tilman, 116
Schwarzschild, Karl, 119, 169
Serbia, 84–86
Shapley, Harlow, 163
Silberstein, Ludwig, 145, 146
Slipher, Vesto M., 160–61
Sobral (Brazil), 126–29, 133–34, 135, 137, 143
Soldner, Johann Georg von, 36
space
  empty, 158, 159
  expansion of, 179
  See also universe
spacetime, 22
  curvature, 59, 65, 94, 96, 118, 143–44
  four dimensions, 32–33, 190
spacetime singularity, 168, 169–71, 213, 218
special theory of relativity
  development, 17, 21–22
  effects of gravity, 27–28, 59

explanation for lack of ether, 24–25
influences, 19–21
mathematics related to, 18, 32–33, 63
redshift, 32
reference frames, 21
spherical geometry, 51–52, 100
curvature of space, 94
of universe, 189, 192–93
Stachel, John, 114, 116
standard candles, 162, 203
stars
bending of light. *See* light deflection
binary pulsars, 201
collapse into black holes, 169
double, 77–78
exploding, 1–6
formation of, 171–72, 177
in Milky Way, 154
neutron, 3–4
nuclear reactions in, 182
Schwarzschild radius, 169
variable, 161–62, 163
Steinhardt, Paul, 11, 197–99, 206
Struve, Hermann, 79–80, 81, 89
Sun, 74–75, 143–44
See also eclipses; light deflection
Supernova Cosmology Project, 195, 199
supernovae
detecting, 215–16
explosions, 184
Type Ia, 2–6, 7, 162, 195, 203–4
Type II, 3–4
superstring theory, 209–10
super-universe, 176

telescopes, 173, 201, 215, 216
tensors, 64–65
curvature, 113–14
energy-momentum, 113, 117, 152, 181
matter, 68
metric, 65, 96, 114, 117, 151–52
Ricci, 114, 115, 117
Thomson, Joseph, 142, 145–46
time
in black holes, 169, 171
evolution of space through, 190, 193–94
topology, 97–99, 212–13, 218
trapped surfaces, 169
Turner, Michael, 204
Turok, Neil, 211, 212–13

unified field theory
cosmologists' interest in, 210–11
Einstein's search for, 100, 160, 211, 219

universe
  age of, 171
  chemical elements in, 177–78
  cosmic background radiation, 175, 176–77
  distribution of matter, 184–87
  Einstein's field equation as model of, 151–52, 156–57
  of empty space, 158, 159
  energy levels in, 207–8
  expansion of. *See* expansion of universe
  "funny energy" (negative pressure or vacuum energy), 10–11, 180, 204–6
  geometry of. *See* geometry of universe
  as infinite, 9, 150–51, 153–54
  lack of center, 179
  mass. *See* mass of universe
  Newton on, 150–51
  origins of, 171, 217–18
  static model of, 156–57
  super-, 176
  temperature of, 176, 177
  See also big bang
Updike, John, 181

variable stars, 161–62, 163
vectors, 33, 65
Voss, Aurel, 114

Wallis, John, 50
Weinberg, Steven, 167, 207
Weyl, Hermann, 165
white holes, 171
Wilhelm II, Kaiser, 85–87
Wilson, Robert, 177
World War I, 84–87, 118–19, 121, 123, 125–26, 162

zeta function, 92
Zürich Notebook (Einstein), 37, 116–17